T0397510

Smith–Purcell Radiation

Smith–Purcell Radiation

Basic Theory and Applications

George Doucas

Great Clarendon Street, Oxford, OX2 6DP,
United Kingdom

Oxford University Press is a department of the University of Oxford.
It furthers the University's objective of excellence in research, scholarship,
and education by publishing worldwide. Oxford is a registered trade mark of
Oxford University Press in the UK and in certain other countries

Published in the United States of America by Oxford University Press
198 Madison Avenue, New York, NY 10016, United States of America

British Library Cataloguing in Publication Data
Data available

Library of Congress Control Number: 2024946955

ISBN 9780198951346

DOI: 10.1093/9780198951360.001.0001

Printed and bound by
CPI Group (UK) Ltd, Croydon, CR0 4YY

The manufacturer's authorised representative in the EU for product safety is Oxford
University Press España S.A. of El Parque Empresarial San Fernando de Henares,
Avenida de Castilla, 2 – 28830 Madrid
(www.oup.es/en or product.safety@oup.com). OUP España S.A. also
acts as importer into Spain of products made by the manufacturer.

For Immy, Kit, Cara, and Gabriella

Preface

The contents of this book are based on work carried out, over a number of years, not only here in Oxford but also at the Technical University Munich (Germany), ENEA at Frascati (Italy), FOM (Netherlands), SLAC (United States), and KEK (Japan). In all these places I have been fortunate to receive valuable advice and help from numerous colleagues, too many to list here. I am grateful to all of them.

However, I would like to put on record my special gratitude to three, late, colleagues: John Walsh of Dartmouth College (United States) who introduced me to the subject of Smith-Purcell radiation; John Mulvey of the University of Oxford who was instrumental in setting up the experiments at Oxford, the first to demonstrate Smith–Purcell radiation at relativistic energies; and Maurice Kimmitt (University of Essex) who managed to teach me the basics of far-infrared techniques. Without their support and advice, this work would not have been possible.

The mathematical formulation of the basic idea that the origin of the radiation is due to currents induced on the surface of the grating was developed in close collaboration with Hayden Brownell, then at Dartmouth College. I am especially grateful to Peter Huggard (STFC Harwell) for his numerous thoughtful comments on the manuscript, especially on issues related to TeraHertz radiation, and to Ivan Konoplev (Culham Laboratory), not only for our close collaboration on some of the experiments but also for many discussions on both the theory and the possible applications of this type of radiation.

Oxford, July 2024

Contents

Introduction

In a very short letter published in 1953 [1], S. Smith and E. Purcell reported the first observation of 'visible light from localized surface charges moving across a grating'. What the paper describes is the fact that when an electron (or any charged particle) passes very close to the surface of a metallic grating, light will be emitted from the grating surface. The phrasing of the text suggests that the experiment was not an incidental observation but, rather, the confirmation of Purcell's expectation. The letter summarized, with remarkable brevity, the main characteristics of the phenomenon:

(a) the wavelength (λ) of the radiation depends on the ratio (β) of the electron velocity to the speed of light (c), on the period (l) of the grating and on the angle of observation (θ) relative to the beam direction, according to the relationship:

$$\lambda = \frac{l}{n}\left(\frac{1}{\beta} - \cos\theta\right) \tag{1}$$

(b) the origin of the radiation is the surface of the grating,
(c) the radiation is highly polarized, and
(d) more than one order of radiation (n) could be observed.

The authors concluded that the observed effect, which has come to be known as Smith–Purcell (SP) radiation, 'might have interesting applications'. The essential feature of this phenomenon is encapsulated in this simple equation, namely that it is possible, at least in principle, to produce a tuneable source of radiation over a broad range of wavelengths by a suitable choice of the other four parameters in (1). In practice, the parameters that are most readily adjustable are the period of the grating and the angle of observation. It is of some interest to note that the paper was published in the year after E. Purcell and F. Bloch shared the Nobel prize 'for their development of new methods for nuclear magnetic precision measurements and discoveries in connection therewith' (according to the Nobel Foundation citation) and that, in spite of Purcell's extensive work on electromagnetism, SP radiation is the only physical process explicitly associated with his name.

Subsequent activity centred on the two basic questions: what is the best theoretical description of the origin of this radiation, and what are its potential uses, if any? Broadly speaking, this activity can be divided into two periods. The first, extending up to about the late 1980s, is characterized on the experimental side by investigations

Smith-Purcell Radiation. George Doucas, Oxford University Press. © George Doucas (2025).
DOI: 10.1093/9780198951360.003.0001

in the visible part of the spectrum, using rather low energy DC beams (typically around 100 keV) and optical gratings with micron or sub-micron periods. Indeed, the original work of Smith and Purcell used a 300 keV beam and a grating with a period of 1.67 μm. A common feature of the early experimental work is that the emphasis is on establishing the basic properties of this radiative process, rather than attempting detailed measurements of the radiated energy and comparisons with a specific theory.

From the early 1990s on, there is a marked change in the way that SP radiation is perceived and investigated. On the experimental side, the beam energies have increased by orders of magnitude and are derived from large accelerator facilities. The experiments at Oxford [2] were the first to be carried out in the relativistic regime (3.6 MeV) and were followed by many others at energies extending up to 28 GeV [3]. The electron beams were now bunched, rather than continuous, while the gratings had periods of tens of microns or even millimetres. This reflects the fact that the perceived uses of this radiative process had also changed, with interest centred on the far infrared part of the spectrum, namely in the wavelength region between 30 and 3000 μm, approximately; this is also referred to in more recent years as the TeraHertz (THz) region (the terms 'far infrared' (FIR) and 'THz' are used interchangeably in this text). This perception was not surprising since the optical region is well populated with sources and SP radiation was considered to be unlikely to have a significant impact in this area. As it turns out, this was a rather pessimistic view of its likely usefulness in the visible part of the spectrum and in the last few years there has been great interest in SP radiation as a tuneable source of radiation in the near infrared (see Chapter 7). The far infrared, on the other hand, is a part of the spectrum that is of great interest to a number of disciplines but has a limited number of widely tuneable sources. SP radiation was bound to attract interest because of its in-built tuneability and its capability to deliver significant levels of radiated energy.

This book does not aim to provide a detailed historical account of the evolution of the subject but, rather, a broad overview of the current state of affairs and an assessment of its potential applications. The emphasis is on the presentation of the basic physical ideas, rather than an extensive and rigorous mathematical analysis. The only exceptions to this statement are Chapters 1 and 3 which discuss the surface currents induced on a metallic grating and the determination of the time profile of ultra-short electron bunches, respectively, and where a more detailed analysis was justified. The book is aimed primarily at experimentalists but it is hoped that the more theoretically minded readers might still find enough gaps or unexplored issues in the theory to attract their interest.

References

1. S.J. Smith and E.M. Purcell, Phys. Rev. **92** (1953) 1069
2. G. Doucas et al, Phys. Rev. Lett. **69** (1992) 1761
3. V. Blackmore et al, Phys Rev. Special Topics Accel. & Beams **12** (2009) 032803

1

Surface Current Theory

Smith–Purcell (SP) radiation is essentially a phenomenon of classical electrodynamics. An electron, or in fact any charged particle, travelling above a grating with perfect surface conductivity will induce a surface charge density on that surface. This 'patch' of surface charge is accelerated as it is being dragged along the grating profile and its shape changes periodically. Since accelerated charges emit radiation, they are the origin of the SP radiation. A large number of conduction electrons are set in motion on the metallic surface, but their displacements must be minute. The surface is assumed to be perfectly conducting, which implies that the electrons can respond instantaneously to any changes in the electric field. This is a basic assumption for the discussion in this chapter and is justified for almost all metals for wavelengths down to a few microns [1]. Although the origin of the radiation is the motion of the surface charges on the grating, this is achieved at the expense of the electron beam energy. This simple physical picture was the explanation suggested in the original paper by Smith and Purcell and is the theoretical description that will be followed in this book. It is a method of analysis which was developed in a number of publications, primarily by the Dartmouth–Oxford groups [2–3] and by others [4], and which has been tested experimentally over a wide range of beam energies, from the few MeV to the multi-GeV region. A brief critique of the alternative theoretical treatments will be given in the final section of this chapter.

The intention here is not only to give an outline of this particular theory but also to allow the reader to calculate the expected yield of SP radiation from any specific experiment. The necessary calculations are semi-analytical but there is enough information for the readers to develop their version of a suitable simulation code. There are a number of assumptions or approximations involved and these are stated clearly in the course of the narrative. The formulae are given in the CGS system of units, although the results of the calculations are given in more familiar units, appropriate to the magnitude of the physical quantity in question (e.g. radiated energy in J, rather than erg, grating periods in mm or μm, rather than m or cm, etc.).

1.1 Surface currents on a blazed grating

In general, the radiated energy per unit solid angle $d\Omega$ and per unit frequency $d\omega$ by a continuous distribution of accelerated charges can be expressed as [5]:

$$\frac{d^2 I}{d\omega d\Omega} = \frac{\omega^2}{4\pi^2 c^3}\left|\int_{-\infty}^{\infty} dt \int \overline{n} \times (\overline{n} \times \overline{J}) \exp[i\omega(t - \frac{\overline{n}.\overline{r}}{c})] dxdydz\right|^2 \tag{1.1}$$

Smith-Purcell Radiation. George Doucas, Oxford University Press. © George Doucas (2025).
DOI: 10.1093/9780198951360.003.0002

where:

$\overline{n} = (\sin\theta\cos\varphi, \sin\theta\sin\varphi, \cos\theta)$ is the unit vector in the direction of observation, $\bar{J}$ is the current density and $\bar{r} = (x, y, z)$ is the position vector of the charge distribution. Although the acceleration of the charges is not shown explicitly in (1.1) and integration over all time, even when the velocity might be constant, is counterintuitive, the expression can be shown to be correct as it stands [5] and relatively easy to apply. The coordinate system used in this chapter is shown in fig. 1.1. Note that in the CGS system the squared modulus of (1.1) has dimensions of (statC.cm)^2.

Consider first the case of a single electron travelling at a height x above the surface of a blazed grating (sometimes referred to as 'echelette' grating) consisting of two facets only, with lengths l_1 and l_2, respectively, period (l) and blaze angles (α_1, α_2) that are complementary, with $|a_1| + |a_2| = 90^0$; one such period is shown schematically in fig. 1.2. Note, however, that in the convention used here angles are measured relative to the beam direction and therefore, if for example in fig. 1.2 $a_1 = 30°$, then $a_2 = -60°$. Alternative profiles could be treated as special cases of a blazed grating. The grating is assumed to have N periods hence its length is $Z = Nl$ and its width is (w).

We are interested in the charge distribution on the surface of the grating. Hence, the coordinate x is a function of z and the induced current density $\bar{J}(J_x, J_y, J_z)$ will now be a 'linear' density (in statC/cm/s). The general expression (1.1) which refers to a charge distribution in 3-D space can now be rewritten for the present 2-D case as:

$$\frac{d^2I}{d\omega d\Omega} = \frac{\omega^2}{4\pi^2c^3}\left|\int_{-\infty}^{\infty} dt \int \overline{n}\times(\overline{n}\times\bar{J})\exp[i\omega(t-\frac{\overline{n}.\bar{r}}{c})]dydz\right|^2 \qquad (1.1a)$$

It is easier to carry out the integrations first and leave the vector multiplication at the end. The result of the triple integration inside the squared modulus of (1.1a) is a complex vector $\overline{G}$, which represents the integrated current density and which has to be evaluated separately for each of the two facets of the grating yielding two quantities ($\overline{G}_1$ and $\overline{G}_2$), before the results can be added and the magnitude of $\overline{G}$ can be determined.

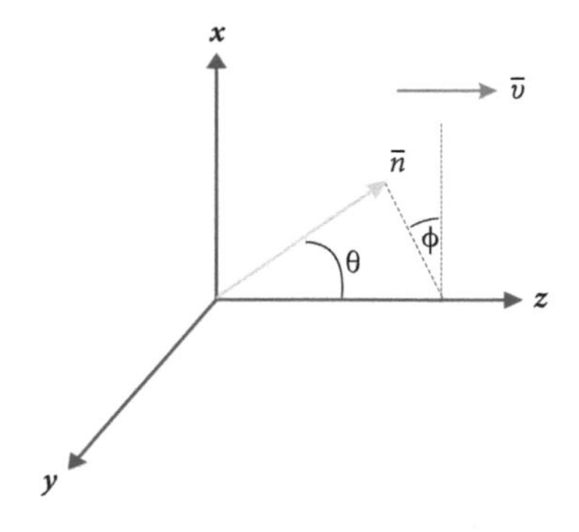

Fig. 1.1 Coordinate system. The charged particle is travelling with velocity υ in the z direction.

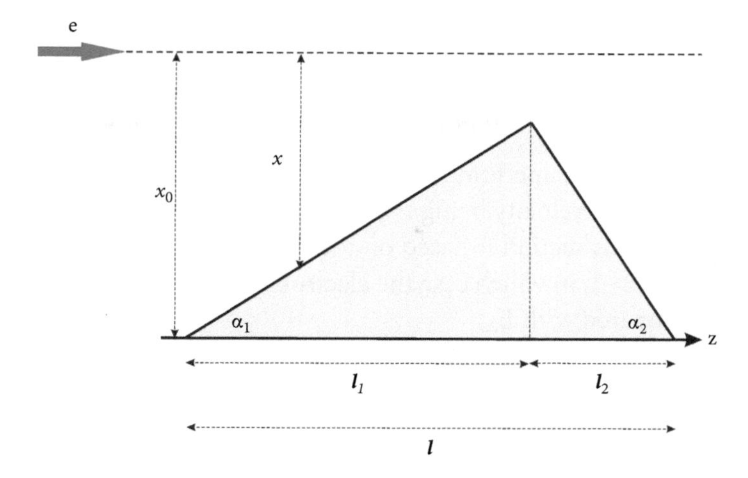

Fig. 1.2 Schematic of one period of a metallic echelette grating.

Vector $\overline{G}$ can then be used to evaluate the vector triple product $\bar{n} \times (\bar{n} \times \bar{G})$ and complete the calculation of the radiated energy.

It is assumed that the electron has a velocity $\bar{\beta}(0, 0, \beta)$ in the z direction, but the induced surface charge will have velocity components (v_x, v_y, v_z) that will depend not only on β but also on the shape of the grating. It is also assumed that v_x and v_y are related and that v_y depends linearly on y and inversely on x. The velocity components are then:

$$v_x = \frac{dx}{dt} = v tan\alpha_j$$

$$v_y = \frac{dy}{dt} = \frac{y}{x} v_x$$

$$v_z = v$$

The surface charge density (σ) induced on the grating surface is given by the usual expression for induced surface charge density on a conducting plane, but taking into account the Lorentz transformed transverse field E_2 of the moving electron (see section 11.10 in ref. [5]):

$$\sigma = \frac{2E_2}{4\pi} = \frac{\gamma e x}{2\pi\left[x^2 + y^2 + \gamma^2 (z - vt)^2\right]^{3/2}} \tag{1.2}$$

where x is the 'impact parameter', i.e. the electron's distance above the grating. Depending on the shape of the grating profile, x can be a variable. Since $\bar{J} = \sigma\bar{v}$, the vector $\bar{J}$ can be written:

$$\bar{J} = \frac{\gamma e x v}{2\pi\left[x^2 + y^2 + \gamma^2 (z - vt)^2\right]^{3/2}} \left(tan\alpha_1, \frac{y}{x} tan\alpha_1, 1\right) \tag{1.2a}$$

It is evident that it would be inaccurate to think that the same surface electrons are being dragged along by the beam, from one end of the grating to the other. A better physical picture would be that of a surface disturbance pattern that is induced by the beam and which changes as it moves up the slope of the facet. The rate of change of this pattern may be superluminal but this should be seen as analogous to the situation of the phase velocity being > c. It should also be noted that (1.2) and the whole analysis of this section is based on the implicit assumption that the beam is relativistic with $\gamma \gg 1$, in which case the electron's longitudinal field component E_1 is negligible compared with E_2.

Returning now to the integrals of (1.1a), it is best to integrate first with respect to time by means of the substitution:

$$u = \frac{\gamma(z - vt)}{\sqrt{x^2 + y^2}}$$

This yields the expression:

$$-\frac{e}{\pi}\frac{x\,\omega\exp[i\omega z/v]}{v\gamma\sqrt{x^2 + y^2}}K_1\left(\frac{\omega}{v\gamma}\sqrt{x^2 + y^2}\right)$$

for the time integral, where K_1 is a modified Bessel function.

Assuming that the beam is travelling along the $y = 0$ line and that, therefore, the charge distribution is symmetrical in the y-direction along the line $y = 0$, it can be shown, after some manipulation of the relevant expressions, that the remaining double integrals over y and z can be expressed as:

$$QU1 \approx (2/\pi)\int_{z_s}^{z_f}\cos(Dz - k_x\chi + \psi)dz\int_0^{w/2}\frac{kx}{\beta\gamma\sqrt{x^2 + y^2}}K_1\left(\frac{k\sqrt{x^2 + y^2}}{\beta\gamma}\right)\cos(k_y y)dy \tag{1.3}$$

$$QU2 \approx (2/\pi)\int_{z_s}^{z_f}\sin(Dz - k_x\chi + \psi)dz\int_0^{w/2}\frac{kx}{\beta\gamma\sqrt{x^2 + y^2}}K_1\left(\frac{k\sqrt{x^2 + y^2}}{\beta\gamma}\right)\cos(k_y y)dy \tag{1.4}$$

$$QU3 \approx (2/\pi)\int_{z_s}^{z_f}\sin(Dz - k_x\chi + \psi)dz\int_0^{w/2}\frac{kx}{\beta\gamma\sqrt{x^2 + y^2}}K_1\left(\frac{k\sqrt{x^2 + y^2}}{\beta\gamma}\right)\sin(k_y y)dy \tag{1.5}$$

$$QU4 \approx -(2/\pi)\int_{z_s}^{z_f}\cos(Dz - k_x\chi + \psi)dz\int_0^{w/2}\frac{kx}{\beta\gamma\sqrt{x^2 + y^2}}K_1\left(\frac{k\sqrt{x^2 + y^2}}{\beta\gamma}\right)\sin(k_y y)dy \tag{1.6}$$

The above expressions are the real and imaginary parts for the z-integration ($QU1$ and $QU2$, respectively) and ditto for the y-integration ($QU3$ and $QU4$). The symbol $\boldsymbol{k}$ is the wave vector with components $k_x = k\sin\theta\cos\varphi$, $k_y = k\sin\theta\sin\varphi$, $k_z = k\cos\theta$. For the general case of an inclined facet, the impact parameter (x) is a function of z and assuming that the base of the grating profile is the reference plane in the x-direction, it can be expressed as:

$$x = x_0 - z\tan\alpha + \chi$$

where α is the blaze angle of the relevant facet (see fig. 1.2). The other quantities that appear in the above expressions are defined as follows:

$$D = k\left(\frac{1}{\beta} - \cos\theta\right) - k_x\tan\alpha$$

- z_s and z_f are the limits for the z-integration and are equal to $(0, l_1)$ for facet 1 and (l_1, l) for facet 2.
- $\chi = 0$ for facet 1 and $\chi = l\tan\alpha_2$ for facet 2.
- ψ is a phase factor that is needed in order to account for the time delay between the emissions from the 1st and 2nd facets. Its value depends on the fraction of the interaction time spent by the particle over each facet, i.e. on the relative lengths of the facets. Therefore, one could either assign zero phase to facet 1, in which case the phase of facet 2 would be $2\pi\frac{l_1}{l}$ or, conversely, assign the value 0 to the 2nd facet and $2\pi\frac{l_2}{l}$ to the 1st.

The above expressions, with the appropriate values for the blaze angle α and for the dummy variables χ and ψ, are applicable to either facet. Note that the y integrand is an even function. In any case, the subsequent procedure is the same: calculate G_{x1}, G_{y1}, G_{z1} for facet 1 (total of six components, since $\overline{G}$ is complex); repeat for facet 2 and then add to obtain G_x, G_y, G_z. Once the magnitude of $\overline{G}$ is known, the single-electron yield of SP radiation can be calculated.

Although at least some of these integrations are necessarily numerical, the computational effort involved will depend on the assumption made about the width (w) of the grating.

1.1a Grating of infinite width

If this assumption is justified, the limits of the y-integration are 0 and ∞ and the evaluation of the integrals (1.3–1.6) is simplified greatly because of the availability of analytical expressions for the y-integrals [6]. The final expression for the one-electron yield from a single period of a grating is:

$$\frac{d^2I}{d\omega d\Omega} = \frac{e^2\omega^2 l^2}{4\pi^2 c^3} e^{-\frac{2x_0}{\lambda_e}} R^2 \tag{1.7}$$

The two other new quantities that appear in the above expressions are the 'evanescent wavelength' λ_e and the dimensionless quantity R^2 which is sometimes referred to, somewhat confusingly, as the 'grating efficiency' or 'radiation factor'. The evanescent wavelength is defined as:

$$\lambda_e = \frac{\beta\gamma\lambda}{2\pi\sqrt{1+\beta^2\gamma^2\sin^2\theta\sin^2\phi}}$$

and is a measure of the coupling efficiency between beam and grating. At short wavelengths λ_e tends to be small and it is necessary to bring the beam close to the grating in order to get appreciable output. One might be tempted to conclude that for a highly relativistic beam ($\gamma \gg 1$) and for detection at $\phi = 0$, λ_e would be large. This, however, is a situation that cannot be realized experimentally since any detector is bound to have a small but finite azimuthal acceptance which would make the second term in the square root $\gg 1$. For such a high energy beam and observation at $\theta = 90°$,

$$\lambda_e \cong \frac{\lambda}{2\pi sin\phi}$$

and the coupling efficiency will depend on the emitted wavelength and on the azimuthal angle ϕ.

The grating efficiency factor R^2 is a complicated expression involving the details of grating profile and has to be evaluated numerically. However, in the case of certain idealized grating profiles it is possible to derive analytical expressions for R^2 (see also sections 1.1c and 1.2 for further comments on this factor). Thus, the overall efficiency of a grating in generating radiation will depend both on the position of the charged particle relative to the grating surface, i.e. on the exponential factor, and also on the details of the design of the grating.

The expression (1.7) refers to the expected yield from a single period. For a grating with N periods, the total grating length is $Z = Nl$ and the situation is analogous to that of a linear array of equal oscillators. If we assume that there is zero phase difference between them, i.e. that the SP condition is satisfied for 1st order emission, then l^2 in (1.7) must be replaced with $(Nl)^2$. For an observer far away from the grating, the resolving power of the grating is given by $d\omega/\omega \cong 1/nN$ [7]; hence, substituting into (1.7):

$$\frac{dI}{d\Omega} = \frac{e^2\omega^3}{4\pi^2c^3}\frac{Zl}{n}e^{-\frac{2x_0}{\lambda_e}}R^2 \tag{1.7a}$$

This is the expression for the SP radiation from a metallic grating of perfect conductivity. Note that the order of emission n appears for the first time in (1.7a) because of the summation over N periods.

1.1b Grating of finite width

If the assumption of infinite width is not appropriate, then the integrals (1.3–1.6) must be evaluated numerically. However, in this case there is no explicit relationship between yield and x_0, as seen in the exponential term of the above eqs (1.7–1.7a). Instead, the dependence on x_0 is now contained within the factor R^2. The corresponding expressions for (1.7, 1.7a) are now:

$$\frac{d^2I}{d\omega d\Omega} = \frac{e^2\omega^2l^2}{4\pi^2c^3}R^2 \tag{1.8}$$

and:

$$\frac{dI}{d\Omega} = \frac{e^2\omega^3}{4\pi^2c^3}\frac{Zl}{n}R^2 \tag{1.8a}$$

Assuming that the detector is located at 'infinity' (this will be discussed later on), then the basic SP expression for wavelength vs observation angle is satisfied and (1.8a) can be rewritten as:

$$\frac{dI}{d\Omega} = 2\pi e^2\frac{Z}{l^2}\frac{n^2\beta^3}{(1-\beta\cos\theta)^3}R^2 \tag{1.8b}$$

The choice of the appropriate limits of the y-integration is really a question about the role of the width of the grating in determining the radiation yield. The best way to answer this is to investigate the *approximate* size of the surface charge density σ, i.e. of the 'footprint' of the electron, on the surface. This can be done by evaluating the expression for the surface charge density (1.2) assuming arbitrarily that at $t = 0$ the electron is at $z = 0$. The results of this calculation, for two values of the relativistic factor γ and two values of the beam-grating separation, are shown as normalized equal density elliptical contours in figs 1.3a–1.3d. Different scales have been used for the z-axes in order to make the plots clearer. In the y direction, i.e. along the width of the grating, the crucial factor is the beam-grating separation x (figs 1.3a, 1.3c). Assuming that the grating width is 20 mm, the charge density at the boundary would be 0.1% of its peak value at the grating centre, if x were equal to 1.0 mm.

However, for $x = 5.0$ mm (figs 1.3b, 1.3d) the charge density at the edge of the grating would still be 10% of its peak value and neglect of the grating's finite width would introduce significant errors in the calculation. As expected, in the ultra-relativistic case ($\gamma = 10^4$) the field lines, and hence the surface charge density contours, are highly compacted in the z-direction (figs 1.3c, 1.3d).

Although the 'footprint' of the electron on the surface of the facet does theoretically extend to infinity, the part that matters in the generation of SP radiation is the surface charge that is in the proximity of the normal to the surface. This can be calculated easily from eq. 1.2. Fig. 1.4 shows the fraction of the total surface charge contained within an elliptical contour whose semi-axes (y_m, y_m/γ) are functions of the electron's height x_0 above the surface of the grating. It can be seen that about 50% of the total surface charge is contained within an ellipse with semi-axes ~ ($2x_0$, $2x_0/\gamma$)

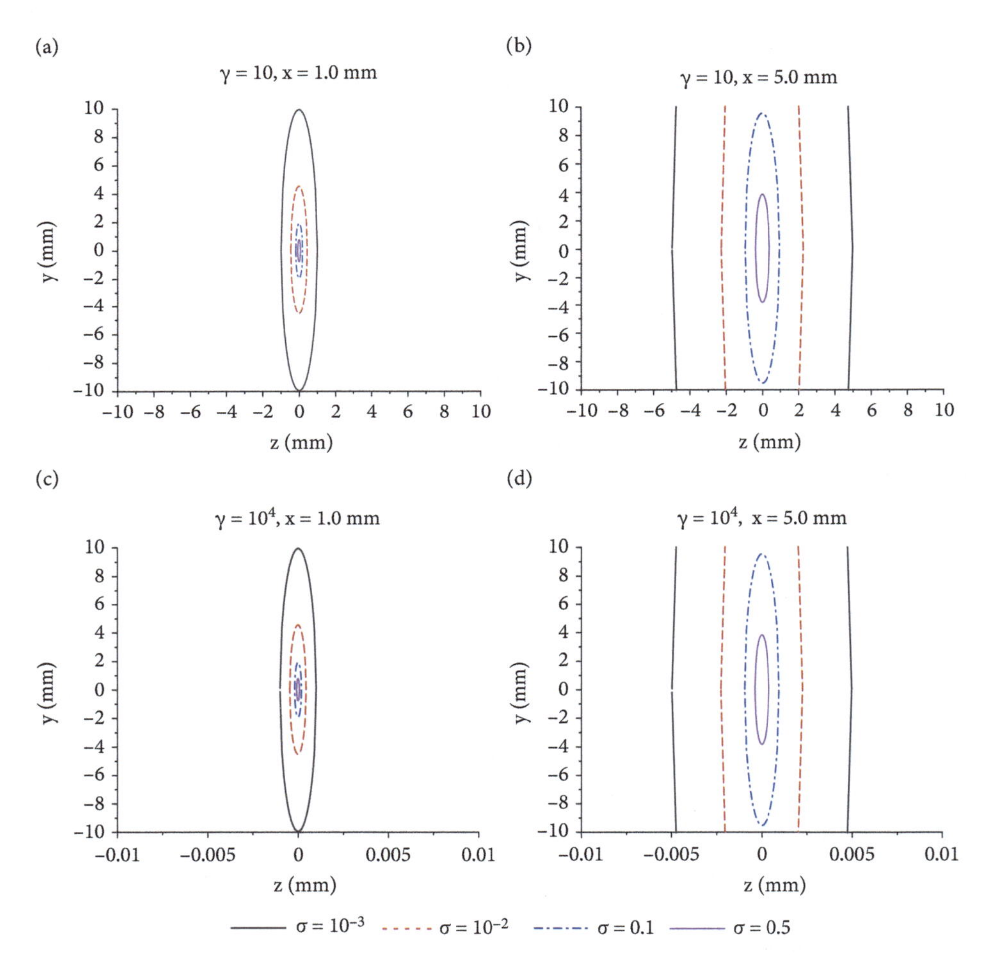

Fig. 1.3 Equal density contours for the surface charge density induced on the surface of a perfectly conducting grating by a relativistic electron having $\gamma = 10$ (a and b) and $\gamma = 10^4$ (c and d) for two different values of electron-grating separation ($x_0 = 1.0$ mm and $x_0 = 5.0$ mm). Note the different horizontal scales.

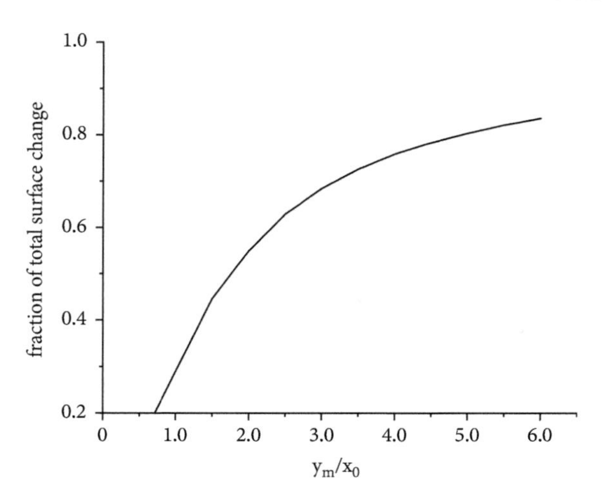

Fig. 1.4 Fraction of the total surface charge contained within an ellipse whose semi-axes (y_m, y_m/γ) are expressed in terms of the electron-grating separation x_0.

in the y and z directions, respectively. This is, therefore, a reasonable approximation to the size of the electron's footprint.

In the case of an echelette grating depicted in fig. 1.2 the size of the footprint would be changing continuously because of the changing value of x and would collapse to a point when $x = 0$, i.e. at the notional intersection of the electron's trajectory with the inclined facet; on the other hand, there would be no change of footprint shape or size in the case of a non-inclined facet.

Since there is no such thing as a grating of 'infinite' width, it is important to estimate the uncertainty introduced in the calculation of the radiated energy. This is shown in fig. 1.5, where the ratio (R) of the single-electron yields obtained from a finite and an infinite width grating is plotted against wavelength. It has been assumed that the grating has a period of 1.0 mm and a blaze angle of 30^0. Three values of γ (10, 100, and 10^4) and three different values of the ratio x_0/w (where w is the grating width) have been considered. The range of wavelengths has been curtailed to emission angles that are likely to be experimentally accessible, i.e. the very forward or backward angles have been neglected. For a rather low energy particle ($\gamma = 10$) the uncertainty introduced by assuming an 'infinite' width is very small ($\ll 10\%$), even for large values of the ratio x_0/w, as shown in fig. 1.5a, although it is evident that it tends to increase at the long wavelength limit. For $\gamma = 100$ (fig. 1.5b) the uncertainties are much more pronounced, especially at the long wavelengths and for large values of x_0/w, when they can be of the order of 40%. This trend becomes even more evident for the ultra-relativistic case of $\gamma = 10^4$ depicted in fig. 1.5c. However, a ratio $x_0/w = 0.2$ would be poor experimental practice indeed and we conclude that for more realistic values of this ratio the uncertainty in this calculation is acceptable,

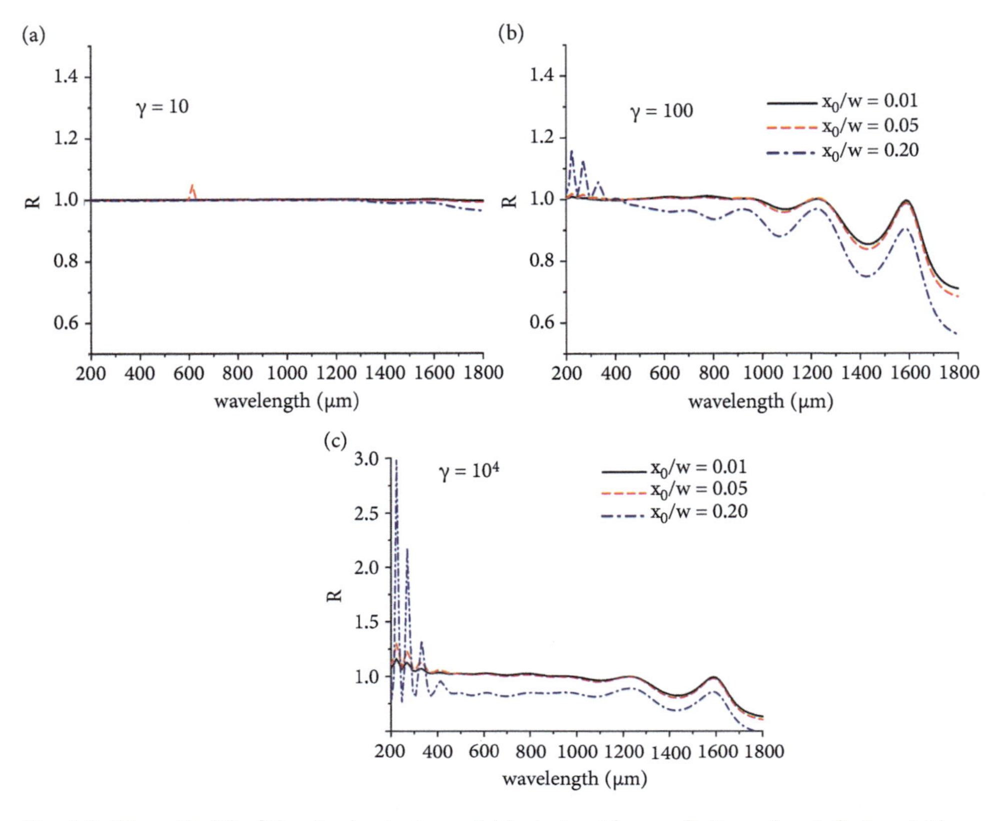

Fig. 1.5 The ratio (R) of the single electron yields derived from a finite and an infinite width grating for three values of the relativistic factor γ: (a) $\gamma = 10$, (b) $\gamma = 100$, and (c) $\gamma = 10^4$. In all plots the three lines correspond to three different ratios x_0/w of the particle position above the grating (x_0) to the grating width (w).

even in the ultra-relativistic case, especially in the context of measurements in the far infrared part of the spectrum. This will be discussed further in a subsequent chapter.

Thus, as long as the grating width is much larger than the beam-grating separation, it makes little difference whether the yield is calculated on the assumption of 'finite' or 'infinite' width. Calculations based on a grating of finite width and the consequent curtailment of the numerical integration to $w/2$, may be justified for small x_0 ($x_0 \ll w/2$) but numerical integration is needed when x_0 does not meet this criterion. There are, however, approximations involved here as well, since any reflections of the surface currents at the discontinuity represented by the grating edge have been neglected. The above discussion can be summarized in a few practical rules: (a) ensure a high-quality beam, well focused over the grating; (b) use as wide a grating as practicable; (c) bring the beam as close to the grating surface as possible, which will also ensure good coupling efficiency as discussed previously.

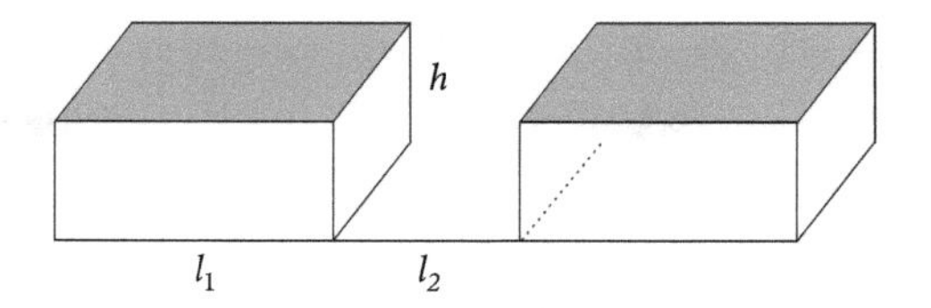

Fig. 1.6 Schematic representation of the lamellar (strip) grating.

1.1c Grating profiles

The preceding discussion was based on a simple profile, consisting of two facets. This is the type of grating that is probably easier to manufacture, especially for operation in the far-infrared, but is not the only possible profile. In fact, one could consider a period that consists of three or more facets but the extra effort in the calculations would be considerable and the benefits unclear. A more realistic alternative profile would be the lamellar grating, shown in schematic form in fig. 1.6.

The period (l) of such a grating is made up of four facets. The two horizontal strips have lengths l_1 and l_2, respectively, and the depth of the groove is h. As a first approximation, the contribution of the vertical facets could be ignored and one would be left with two horizontal strips, which can be treated as described in the previous section. As a further approximation, the contribution of the lower facet, of length l_2, could also be ignored, since its coupling efficiency will be lower, and then the only source of radiation left would be the upper facet. This idealized grating would then consist of thin, co-planar strips, separated by air gaps, and is sometimes referred to as a 'strip' grating. There is an analytic expression for the efficiency factor R^2 of such a grating given by [2, 3]:

$$R^2 = sin^2\theta sin^2\left(\frac{n\pi l_1}{l}\right) \tag{1.9}$$

Reflections that are taking place inside the groove have also been neglected in this treatment. Note that when $l_1 = l$ then $R^2 = 0$, i.e. no SP radiation is emitted from a continuous planar surface. Strip gratings have interesting properties when it comes to the polarization of the radiation, but this will be discussed in a subsequent chapter.

1.2 General comments on the theories of SP radiation

The description of SP radiation in terms of surface currents is not the only possible theoretical treatment. Neither, for that matter, are surface currents only applicable to this particular problem since they could also be applied to diffraction or transition radiation [8]. Since a detailed account of the various alternative theories would be beyond the scope of this book, this section will only present the basic physical principles upon which these theories are based, together with some important references where the reader can obtain further details.

An alternative approach, which received a lot of attention in the early years, considers the field of the travelling electron as a superposition of an infinite set of evanescent waves; when these are diffracted by the grating they can become propagating waves and thus give rise to the observed radiation. This is a plausible approach but the analysis is very mathematical and rather lacking in physical transparency or intuition. The idea was originally suggested by Toraldo di Francia [9] and was further developed by van den Berg [10–11], Haeberle et al [12–13], and others [14–15]. The expressions for the radiated energy (or power) derived by Toraldo di Francia or by Haeberle are very similar to eqs (1.7) or (1.7a) above. The basic difference lies in the evaluation of the grating efficiency, or radiation, factor R^2 (denoted as δ^2 in ref. [2]). The prediction of this theory is that R^2, and hence the SP yield, will *decrease* by orders of magnitude with increasing beam energy, approximately according to a γ^{-2} factor [15]. This is in total disagreement with the predictions of the surface current theory which expects the radiation factor to remain essentially unchanged or to increase with beam energy. Fig. 1.7 is an illustration of this issue for the case of a strip grating of 'infinite' width: fig. 1.7a is taken from reference [15], while fig. 1.7b shows the grating efficiency (averaged over the $-3^0 < \phi < 3^0$ range) calculated according to surface currents. The same grating parameters have been used in both cases, namely l = 1.0 mm, l_1 = 0.5 mm, h = 0.495 mm, but in fig. 1.7a the abscissa is the dimensionless quantity wavelength/period while in fig. 1.7b it is just wavelength; the electron was assumed to be 1.0 mm above the grating surface. The important point in the comparison of these plots is not so much the actual values of the grating efficiency but the expected behaviour with increasing γ.

According to the diffracted wave theory, the grating factor would decrease by four to five orders of magnitude as the beam energy increases from E = 3.5 MeV to 855 MeV, whereas the surface current theory anticipates an increase of about one order of magnitude over the same energy range; in fact, according to the latter theory, there is hardly any change in grating efficiency between 180 MeV ($\gamma \cong 360$) and 855 MeV ($\gamma \cong 1710$). It is also interesting to note that at low energies both theories predict radiation factors (or 'grating efficiencies') of the same order of magnitude; the differences begin to appear for energies higher than about 10 MeV. Reasonable agreement has been claimed between the diffracted wave theory and the results of an experiment carried out at 855 MeV [16]. However, the experiment was carried out in the optical wavelength part of the spectrum where the assumption of perfect conductivity is not valid and the surface current model would not be applicable.

A variation on the surface current idea, but with a much more complicated mathematical treatment, was pursued by the MIT group [17]. Assuming that the grating is a perfect conductor and hence that the tangential component of the incident electric field is equal to that of the scattered field, one can calculate the surface current density J; from the current density it is then possible to derive the vector potential

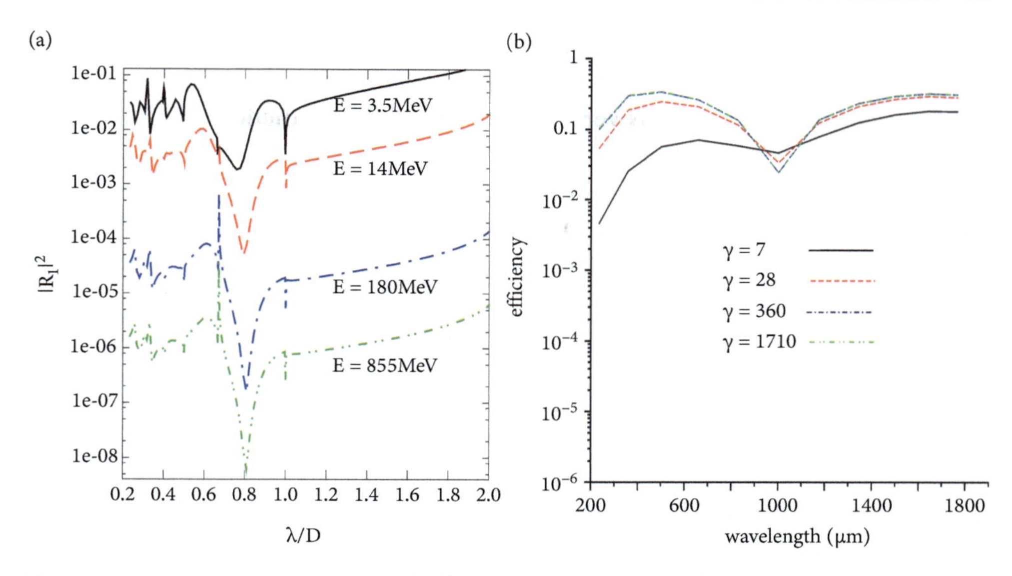

Fig. 1.7 Comparison of the calculated efficiency of a strip grating having a period $l = 1.0$ mm, at different values of the relativistic factor γ. The plots in (a) are taken from ref. [15] and are based on the diffracted wave theory while those of (b) are derived from the surface current theory (the plots for the two highest values of γ in (b) are indistinguishable).

$\overline{A}$ and thence the magnetic component $\overline{H}$ of the far field; this then allows the calculation of the power spectrum. In these calculations, the surface current density appears as part of the integrand and the procedure is referred to as the Electric Field Integral Equation (EFIE) [18]. Very good agreement between this theory and experiments carried out at 15 MeV was reported [19]. A comparison between the surface current (SC) model outlined in this chapter and the EFIE method at GeV energies indicated that they are in broad, but not total, agreement with the EFIE predicting lower outputs by a factor of about 5 [20].

The group in Tomsk, Russia, have also invested considerable effort in the theoretical study of SP radiation. Their work draws attention to the similarities between the SP and diffraction/transition radiation mechanisms and describes the former as 'resonant diffraction radiation'. The formulae that they derive refer primarily to strip gratings (see fig. 1.6). Details of the calculations and further references can be found in [21, 22]. Apart from the above semi-analytical approaches to the problem of SP radiation, there is also the possibility of using various software packages for electromagnetic simulations, like VSim or CST Particle Studio. At present, most of them can calculate accurately the radiation field in the near zone but it is still a challenge to derive accurate information for the far-field zone.

The conclusion from this brief overview of the various theoretical descriptions of SP radiation is that although this is clearly a process of classical electrodynamics, there is no universally accepted theoretical description. The various theories may appear to be ahead of the experimental evidence primarily because precision measurements in the far-infrared, which is an obvious area of interest for this radiative process and where the assumption of perfect conductivity is justified, are not currently achievable. It is thus very difficult to test the predictions of the different theories against experimental data. The surface current theory has been tested over the widest range of energies and agrees with experimental observations to within a factor of 5: this statement will be examined in greater detail in subsequent chapters. It is likely that a similar level of agreement between theory and measurement may be achievable with the EFIE method but it is unfortunate that, so far, there has not been a single experiment that has been analysed by two alternative methods.

1.3 Polarization of SP radiation

In analogy with other closely related radiative processes (bremsstrahlung, transition, Cherenkov etc.), SP radiation must be polarized in directions defined by the vector integral of eq. (1.1a), which is repeated below:

$$\frac{d^2I}{d\omega d\Omega} = \frac{\omega^2}{4\pi^2c^3}\left|\int_{-\infty}^{\infty} dt \int \overline{n} \times (\overline{n} \times \overline{J}) \exp[i\omega(t - \frac{\overline{n}.\overline{r}}{c})]dydz\right|^2$$

Since the direction of the surface current density $\overline{J}$ depends on the grating profile and is not known, it is reasonable to specify the polarization of the radiation with respect to the plane containing the beam direction $\bar{z}$ and the direction of observation $\bar{n}$ (see fig. 1.8); the unit vectors $\overline{\varepsilon_1}$ and $\overline{\varepsilon_2}$ are parallel and perpendicular to this plane, respectively, and both are perpendicular to the unit vector $\bar{n}$ along the observation direction.

As mentioned in section 1.1, the result of the triple integration over t, y, and z is a complex vector $\overline{G}$ which leads to the evaluation of $\overline{n} \times (\overline{n} \times \overline{G}) = (\overline{n}.\overline{G})\overline{n} - \overline{G}$ in the integral of (1.1a). The intensities of the radiation polarized along the directions $\overline{\varepsilon_1}$ and $\overline{\varepsilon_2}$ are I_1 and I_2, respectively; these can be found by taking the scalar product of the above expression with the relevant unit vector and then squaring. However,

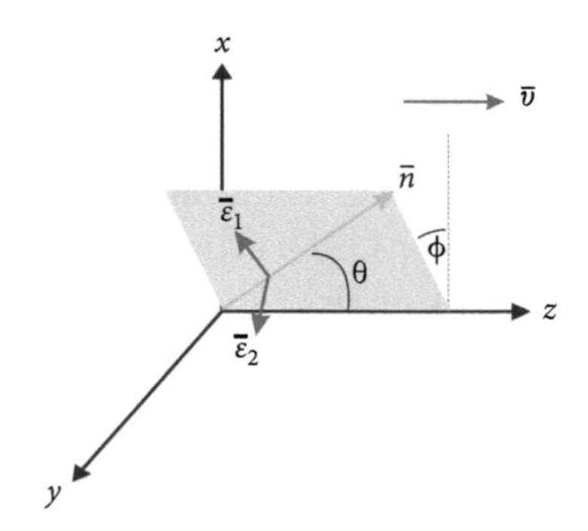

Fig. 1.8 Definition of the two polarization directions $\overline{\varepsilon_1}$ and $\overline{\varepsilon_2}$.

in order to evaluate the projections of this vector on $\overline{\varepsilon_1}$ and $\overline{\varepsilon_2}$, which are $-\overline{\varepsilon_1}.\bar{G}$ and $-\overline{\varepsilon_2}.\bar{G}$, respectively, it is necessary to convert the unit vectors to complex form by 'padding' them with zeros for their imaginary parts. Hence:

$$\overline{\varepsilon_1} = (\cos\theta\cos\phi, 0, \cos\theta\sin\phi, 0, -\sin\theta, 0)$$

$$\overline{\varepsilon_2} = (-sin\phi, 0, \cos\phi, 0, \ 0, 0)$$

In general, the evaluation of the radiation intensities along these two polarization directions is really the evaluation of the grating efficiency factor R^2, or more accurately, the evaluation of *two* efficiency factors R_1^2 and R_2^2, one for each polarization direction. This is best done numerically because the analytic calculation is very long-winded. However, it can be shown that for a grating of infinite width and for the azimuthal angle $\phi = 0$, the radiation is completely polarized in the $\overline{\varepsilon_1}$ direction; hence, the degree of polarization p which is defined as:

$$p = \frac{I_1 - I_2}{I_1 + I_2}$$

becomes equal to 1. This can be understood in purely physical terms by the following argument. At $\phi = 0^0$ the acceleration of the surface charges must be along the beam direction, irrespective of the inclination of the grating facet relative to the beam direction. If the facet is parallel to the beam (strip grating), evidently so, but even when there is a non-zero blaze angle, charge acceleration along the other directions must cancel out for reasons of symmetry. Since the radiation from an accelerated charge is polarized in the plane defined by the acceleration and observation directions [5], the polarization in this limiting case must be in the *x-z* plane. For a blazed grating, there is acceleration of surface charges in other directions as well, apart from that defined by the beam, and the degree of polarization (DOP) varies quite rapidly with the azimuthal and emission angles, the blaze angle of the grating and the order of emission, as shown in the simulations of fig. 1.9 which were carried out on the assumption of a grating of infinite width. The first question, of course, is whether the 'infinite' width is the appropriate assumption and whether one might get different answers from calculations based on the finite width of the grating. The short answer to this is that as long as the wavelength of the radiation is very small compared with the width of the grating, the two types of calculation yield essentially the same answers. Significant differences will start to appear in the backward directions (i.e. longer wavelengths) and/or with gratings of long periods where, again, the wavelengths are less likely to be trivially short, compared with the grating width. It pays to work with gratings that are as wide and as short-period as practicable; one can then take advantage of the far quicker calculations afforded by the assumption of infinite width. In fig. 1.9a the grating has a period of 1.0 mm and a blaze angle of 10^0, while in fig. 1.9b the blaze angle is 30^0; in both cases the beam was assumed to have $\gamma = 100$. The plots show the average DOP for emission orders 1, 2,

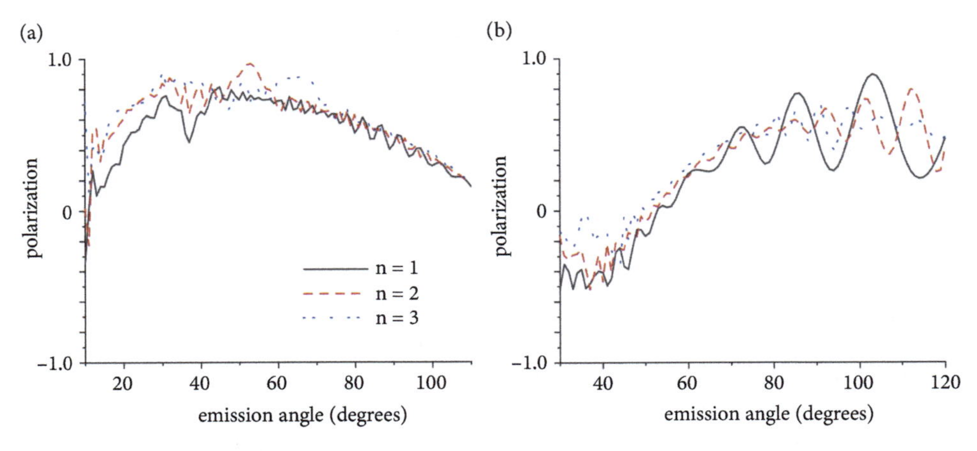

Fig. 1.9 The degree of polarization of the radiation originating from a grating with period = 1.0 mm, as a function of emission angle (θ) for orders 1, 2, and 3; the blaze angle of the grating is 10^0 in (a) and 30^0 in (b). The polarization has been averaged over the range $\theta \pm 3^0$ and $\phi \pm 3^0$.

and 3 and they cover the emission range that is appropriate to each grating, bearing in mind that the predictions of the surface current theory for a metallic grating are meaningful only in the emission range $\alpha < \theta < 90 + \alpha$, where α is the blaze angle of the echelette grating (radiation emitted outside this angular range would have to travel through the metal before it can reach the detector). In any experimental arrangement there will be a finite acceptance in the θ, ϕ directions determined by the characteristics of the optical system; hence, the degree of polarization has been averaged over the (plausible) acceptance range $\theta \pm 3^0$ and $\phi \pm 3^0$.

Although there are differences in the DOP between the different emission orders, the general trend is the same but the oscillations are less noticeable in orders 2 and 3. In both cases the DOP is close to zero, or even negative, around the blaze angle but rises to positive values thereafter, before beginning to decrease again. In these simulations the grating with the low blaze angle shows a smoother variation of DOP with emission angle and higher values of DOP at low emission angles. This is not unexpected since a low blaze facet is closer to the idea of a lamellar grating where, as mentioned earlier, the DOP is predicted to be close to unity, irrespective of the azimuthal angle ϕ. In summary, it would be accurate to say that SP radiation is linearly polarized, predominantly in the plane defined by the beam direction and the direction of observation, but beyond that it is not possible to make any further quantitative comments. The DOP will depend strongly on the profile of the grating and will have to be calculated for each specific grating.

Unfortunately, there are very few comparisons between theory and experiment on this particular topic [14,23,24] and these are inconclusive, either because the comparison was done at one wavelength only and the agreement was modest [14] or because the experimental uncertainties in the reported measurements are high and hence, they do not lead to any firm conclusions [23, 24]. This is regrettable because accurate knowledge of the DOP of SP radiation would be an excellent method of discriminating against other sources of background radiation, as will be explained in a subsequent chapter.

1.4 Some comments on the intensity of SP radiation

How does the strength of SP radiation compare with that of other, similar, radiative processes? There is no unique answer to that, of course, because the answer depends amongst other parameters, on the profile of the grating and on the coupling between beam and grating. Nevertheless, it is of some interest to attempt an order-of-magnitude comparison with transition radiation (TR) which is a well understood process. As far as TR is concerned, the expression for the *total* energy (I) radiated per electron and per single interface of a foil of infinite width has already been derived [25, 26]:

$$I = \frac{\gamma\hbar\omega_p}{3\,(137)} \cong \frac{\gamma\hbar\omega_p}{400} \;(eV)$$

The quantity ω_p is the plasma frequency of the metallic foil; typically, $\omega_p \cong 10^{15}s^{-1}$ and $\hbar\omega_p$ for most metals is of the order of a few eV. According to [25] about 50% of the TR energy is emitted in the frequency range $0.1\gamma\omega_p<\omega<\gamma\omega_p$. This means that even for a relatively low γ of about 10, the emitted energy will involve frequencies that are far higher than those produced in a typical SP experiment (usually of the order of $\omega \cong 10^{13}s^{-1}$). Therefore, a more meaningful comparison would be not with the total TR yield but with the TR yield in the part of the spectrum that overlaps that of a typical SP experiment.

In order to calculate the SP yield in emission order 1, again for a single electron, one can assume the following parameters which are based on past experiments and are appropriate for this investigation:

1. Beam energy, in terms of γ: 10, 100, and 4×10^4
2. Grating period: 0.5 mm
3. Grating length: 36 mm
4. Grating width: 20 mm
5. Grating type: echelette, with a 30^0 blaze angle.
6. Electron-grating separation: 0.5 mm
7. Azimuthal range: -9^0 to $+9^0$, in nine 1^0 steps in each direction.
8. Emission angle (θ) range: 30–170^0. Therefore, for this specific grating, the comparison is over the wavelength range between 70 and 1000 μm, approximately.

When estimating the TR yield, it is important to consider also the effect of the finite size of any realistic TR screen. This problem has been dealt with by a number of

Table 1.1 An approximate comparison of the intensities of transition and SP radiations arising from a single electron

	TR screen radius *a* (mm)					SP energy (J)
	5	10	50	500	4×10^{4}	0.5 mm grating
	Radiated energy (J)					
Electron energy (γ)						
10	2.0×10^{-22}	2.0×10^{-22}	2.0×10^{-22}	2.0×10^{-22}	2.0×10^{-22}	4.7×10^{-24}
100	1.8×10^{-22}	1.9×10^{-22}	1.9×10^{-22}	1.9×10^{-22}	1.9×10^{-22}	3.2×10^{-23}
40,000	3.6×10^{-29}	5.8×10^{-28}	3.4×10^{-25}	1.1×10^{-22}	1.9×10^{-22}	3.9×10^{-23}

authors [27, 28]. Following the analysis of [28], the transition radiation yield from a screen of radius (a) is given by:

$$\frac{d^2I}{d\omega d\Omega} = \frac{e^2}{\pi^2 c}\frac{\beta^2 sin^2\theta}{(1-\beta^2 cos^2\theta)^2}[1 - T(\gamma, \omega\alpha, \theta)]^2$$

where T is a correction factor arising from the finite size of the screen and ω is the frequency of the radiation; T is a combination of Bessel functions and is equal to zero for an infinitely wide screen. Since the solid angle $d\Omega = 2\pi sin\theta d\theta$, the above expression can be rewritten in terms of wavelength λ and emission angle θ as follows:

$$\frac{d^2I}{d\lambda d\theta} = \frac{4e^2}{\lambda^2}\frac{\beta^2 sin^3\theta}{(1-\beta^2 cos^2\theta)^2}[1 - T(\gamma, 2\pi\alpha/\lambda, \theta)]^2$$

The integration over θ and λ has to be carried out numerically. The results of the calculation are summarized in Table 1.1.

It is important to emphasize that the calculation of the SP yield is an underestimate not only because it is not possible to calculate accurately over the whole of the hemisphere above the grating but also because the intensity of SP radiation that goes into high order emission has also been neglected. As will be discussed in Chapter 3, higher-order emission is not necessarily negligible. Using purely physical arguments, one might expect TR to be stronger than SP radiation due to the fact that in SP only a fraction of the electron's field lines interacts with the radiating medium (the grating) while in TR all the field lines play a role in the generation of the radiation, assuming that the screen has a large enough radius. However, the greatest part of the emitted TR energy lies in wavelength regions that are far shorter than those that are likely to be practicable for the type of SP experiment considered in this section. It is still possible for TR to exceed SP radiation in the sub-millimetre wavelength region where the two processes might overlap, but anything that suppresses the 'long' wavelengths of TR, such as small screen size and/or very high beam energies will tip the balance in favour of SP radiation.

References

1. M. Born and E. Wolf, 'Principles of Optics', Ch. XIII, Cambridge University Press (2019)
2. J.H. Brownell, J.E. Walsh, and G. Doucas, Phys. Rev. E **57** (1998) 1075
3. S.R. Trotz et al, Phys. Rev. E **61** (2000) 7057
4. J.C. MacDaniel et al, Applied Optics **28** (1989) 4924
5. J.D. Jackson, 'Classical Electrodynamics', John Wiley & Sons, 2nd Edition, Ch. 14, Section 5 (1975)
6. I.S. Gradshteyn and I.M. Ryzhik, 'Table of Integrals, Series and Products', Academic Press (1980)
7. M. Born and E. Wolf, 'Principles of Optics', Ch. VIII, Cambridge University Press (2019)
8. D. V. Karlovets and A. P. Potylitsyn Physics Letters A **373** (2009) 1988
9. G. Toraldo di Francia, Il Nuovo Cimento **XVI** (1960) 61
10. P.M. van den Berg, J. Opt. Soc. America **63** (1973) 689
11. P.M. van den Berg, J. Opt. Soc. America **63** (1973) 1588
12. O. Haeberle et al, Phys. Rev. E **49** (1994) 3340
13. O. Haeberle, Opt. Communications **141** (1997) 237
14. Y. Shibata et al, Phys. Rev. E **57** (1998) 1061
15. G. Kube, Nucl. Instr. and Methods in Phys. Res. B **227** (2005) 180
16. G. Kube et al, Phys. Rev. E **65** (2002) 056501
17. A. Kesar, Phys. Rev. Accel. & Beams **13** (2010) 022804
18. C.A. Balanis, 'Advanced Engineering Electromagnetics', John Wiley & Sons (1989)
19. A. Kesar, R.A. Marsh, and R.J. Temkin Phys. Rev. Accel. & Beams **9** (2006) 022801
20. A. Kesar, private communication.
21. A.P. Potylitsyn, Physics Letters **A 238** (1998) 112
22. A.P. Potylitsyn et al, 'Diffraction Radiation from Relativistic Particles', Springer (2010)
23. H.L. Andrews et al, Phys. Rev. Accel. & Beams **17** (2014) 052802
24. H. Harrison, D.Phil. Thesis, University of Oxford, 2018
25. J.D. Jackson, 'Classical Electrodynamics', 2nd Edition, Ch. 14, Section 9 (1975)
26. M. Cherry et al, Phys. Rev. D **10** (1974) 3594
27. N. Shulga and S. Dobrovol'skii, JETP Lett. **65** (1997) 611
28. S. Casalbuoni et al, Phys. Rev. Accelerators and Beams **12** (2009) 030705

2
Coherence, Monochromaticity, and High Order Emission

The discussion in the preceding chapter dealt with the case of a single electron interacting with a metallic grating of perfect conductivity. From the theoretical point of view, this is the most difficult part of the problem. The present chapter considers the case of a continuous beam or, more realistically, of a bunched beam where one has to consider the superposition of the contributions of the individual electrons in the bunch. This is really the well-known problem of interference [1] which in the context of this discussion leads to the concept of coherence [2]. Its application to the case of SP radiation, however, has some interesting implications which will be explored in some detail later on.

2.1 Coherence

We consider a bunch of N electrons travelling above a grating and we want to investigate how the contributions of these N electrons would add together. The general answer is, of course, that this will depend on the relative positions (x, y, z) of these electrons above the grating; note that in this section the coordinates (x, y, z) refer to the position of the beam particles in the 3-D space above the grating. It is best to start with the familiar problem of the intensity generated by N oscillators, each of unit amplitude, but having different phases ϕ_k. The result is given by:

$$\left|\sum_{k=1}^{N} e^{-i\phi_k}\right|^2 = \sum_{k=1}^{N} e^{-i\phi_k} \sum_{k=1}^{N} e^{i\phi_k} = N + \sum_{k=1}^{N} e^{-i\phi_k} \sum_{j\neq k}^{N} e^{i\phi_j}$$

It is easy to show that for $\phi_j \cong \phi_k$, this becomes $\cong N + N(N-1) \cong N^2$, if $N >> 1$. Next consider the case where the oscillators have unequal amplitudes, in which case the sum is of the form:

$$\left|\sum_{k=1}^{N} e^{-\alpha x_k} e^{-i\phi_k}\right|^2 \equiv \sum_{1}^{N} e^{-\alpha x_k} e^{-i\phi_k} \sum_{1}^{N} e^{-\alpha x_k} e^{i\phi_k} = \sum_{1}^{N} e^{-2\alpha x_k} + \sum_{k=1}^{N} e^{-(\alpha x_k + i\phi_k)} \sum_{j\neq k}^{N} e^{-(\alpha x_j - i\phi_j)}$$

In our specific problem, the 'amplitude' is determined by the strength of coupling to the grating, i.e. the x position of the particle and the phase ϕ_j by its y and z (or t)

Smith-Purcell Radiation. George Doucas, Oxford University Press. © George Doucas (2025).
DOI: 10.1093/9780198951360.003.0003

positions in the bunch (see also 1.1 and 1.7); hence:

$$\phi_j = k_y y_j + \frac{\omega z_j}{v} = k_y y_j + \frac{k z_j}{\beta}$$

When $N \gg 1$ the above sums become integrals over the $x, y, z(t)$ distribution of the particles inside the bunch. If we assume that the shape of the bunch can be described by three un-correlated and normalized distributions *X(x)*, *Y(y)*, *Z(z)* (or *T(t)*), then the integrals are separable. If we make the further, plausible, assumption that the *X(x)* and *Y(y)* distributions are Gaussian centred around (x_0, y_0), then the first summation is replaced by *N* multiplied by the 'incoherent' integral S_{inc}:

$$S_{inc} = \frac{1}{\sigma_x \sqrt{2\pi}} \int_0^{\infty} e^{-\frac{2x}{\lambda_e}} e^{-\frac{(x-x_0)^2}{2\sigma_x^2}} dx \tag{2.1}$$

while the double summation is replaced by N^2 multiplied by the square of a triple integral (the 'coherent' integral) S_{coh}:

$$S_{coh} = \left| \frac{1}{\sigma_x \sqrt{2\pi}} \int_0^{\infty} e^{-\frac{x}{\lambda_e}} e^{-\frac{(x-x_0)^2}{2\sigma_x^2}} dx \right|^2 \left| \frac{1}{\sigma_y \sqrt{2\pi}} \int_{-\infty}^{\infty} e^{-ik_y y} e^{-\frac{(y-y_0)^2}{2\sigma_y^2}} dy \right|^2 \left| \int_0^{\infty} e^{-i\omega t}\, T(t)\, dt \right|^2 \tag{2.2}$$

In the above expression, integration along the *z*-direction has been replaced by integration over time but the distribution *T(t)* of the particles in the time domain is as yet unspecified; note, however, that the integral in the last squared modulus in (2.2) is the Fourier transform of *T(t)*.

Therefore, the SP radiation yield per solid angle from a bunch containing N_e electrons, with $N_e \gg 1$, will be given by the expression:

$$\left(\frac{dI}{d\Omega}\right)_{N_e} = \left(\frac{dI}{d\Omega}\right)_1 \left(N_e S_{inc} + N_e^2 S_{coh}\right) \tag{2.3}$$

where $\left(\frac{dI}{d\Omega}\right)_{Ne}$ is the yield per solid angle from all the N_e particles in the bunch and $\left(\frac{dI}{d\Omega}\right)_1$ is the single-electron yield per solid angle, calculated by any of the theories outlined in the previous chapter, e.g. eq. (1.7a) or eq. (1.8a).

Equations (2.2) and (2.3) lead to the following conclusions:

(a) If the coherent integral is of order unity, then the first term in the parenthesis of (2.3) is negligible compared with the second and can be ignored; the yield from the bunch is then just N_e^2 greater than that from a single electron. This is the regime of coherent emission of radiation. The maximum value for each of the three terms in (2.2) is one, therefore for coherent emission to happen, the squared moduli in the coherent integral S_{coh} must be of order unity. Physically, this means that the bunch must be 'small' and since in the context of this discussion the wavelength is the measure of everything, 'small' means small compared with the wavelength; the contributions of all the electrons in the bunch are then more or less in phase and the bunch appears as a 'lump' of charge. Although not immediately obvious from the above equations, the onset of coherence has a profound effect on the spectral distribution of the emitted radiation. This can be understood by a simple, qualitative argument: the wavelengths that are long compared with the bunch length will be enhanced by coherence while the shorter ones will be largely unaffected. Hence, while the incoherent emission, at least in the relativistic regime, is peaked in the forward direction (shorter wavelengths), the opposite is true for coherent emission. This is shown in fig. 2.1 for the case of a beam with $\gamma = 100$ passing at a height $x_0 = 1.0$ mm over a grating having a period of $l = 1.0$ mm. For the calculation of the incoherent emission (black solid line) it has been assumed that the beam is continuous, with I = 1nA ≅ 6×10^9 e/s; in the coherent example (red dashed line) 6×10^9 electrons have been assumed to be compressed into a Gaussian bunch with a full width at half maximum (FWHM) = 1 ps. The bunched yield is not only orders of magnitude higher than that of the incoherent emission (note the different scales of the vertical axes) but it has also been shifted to longer wavelengths.

(b) As far as the two transverse dimensions (x, y) are concerned, coherence can usually be achieved by bringing the beam to a focus over the centre of the grating.

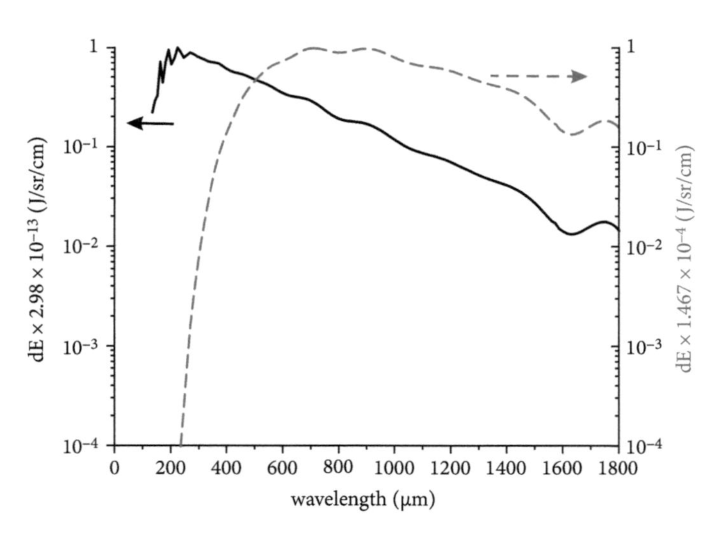

Fig. 2.1 Comparison of the spectral yield of SP radiation from a DC beam (black/solid line) and from a bunched beam (red/dashed line), both having $\gamma = 100$. The DC beam has a current I = 1nA while in the bunched case the 6×10^9 electrons are assumed to be compressed into a 1 ps FWHM Gaussian bunch.

(c) When it comes to the third dimension (z or t), it cannot be assumed *a priori* that the last bracket in eq. (2.2) will be of order unity since this will depend on the time profile $T(t)$ of the bunch. In fact, for a DC beam or a beam compressed into a 'long' bunch, its value will be of order zero and the coherent integral will also be zero. This is referred to as incoherent emission in which case the radiated energy is just N_e times that of a single electron. The magnitude of the Fourier transform of the time profile of a bunch is denoted in this work by the symbol ρ and could be calculated, if the profile were known beforehand. One might ask whether, for a given bunch length (and this is a quantity which is reasonably well known in most cases), there would be any significant changes to ρ depending on the detailed distribution of the charges inside this bunch. Fig. 2.2 is a demonstration of how ρ would vary with frequency in the case of a few plausible time distributions. The curves marked with $\alpha = 1$, 3, or 5 refer to Gaussian bunches that all have a FWHM of 50 fs. They differ in that they are either symmetrical ($\alpha = 1$) or the trailing half of the bunch has a σ that is 3 or 5 times, respectively, greater than that of the leading half. Also shown is the ρ of a 'spike' bunch consisting of a 50 fs FWHM leading edge followed by a 'tail' with a FWHM = 500 fs. Note that at very low frequencies, i.e. long wavelengths, $\rho \to 1$ and there is full coherence, as expected. The answer is that ρ does indeed change with the details of the distribution but by less than an order of magnitude.

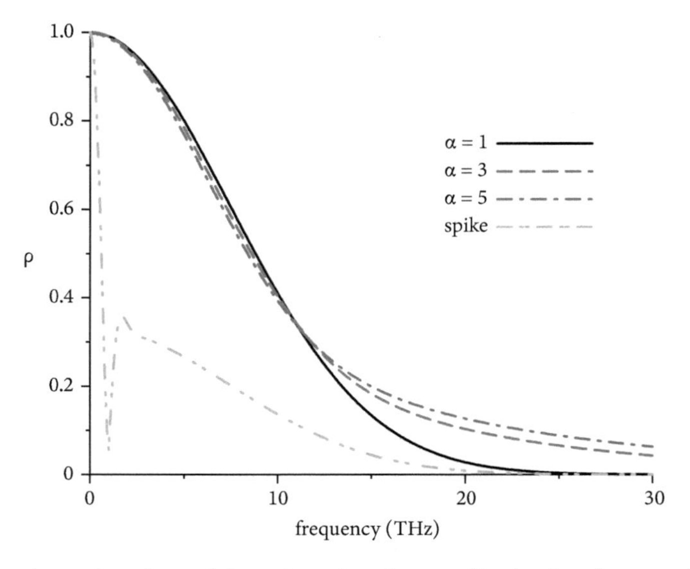

Fig. 2.2 Comparison of ρ values of three Gaussian time profiles having the same FWHM = 50 fs but with different degrees of asymmetry (α) between the leading and trailing parts of the bunch; also shown is the case of a 'spike' followed by a much longer trailing part. See text for further details.

Thus, the details of the charge distribution are encoded in the last bracket of (2.2), i.e. in the values of ρ. In reality, the experimenter is interested in the reverse problem: how to determine the details of the bunch shape from measurements of the radiated energy at different wavelengths. This is an important application of coherent SP radiation which will be discussed in detail in Chapters 4 and 5.

2.2 Far- and near-field measurements

The previous section dealt with the interference of the contributions of the N_e electrons in a bunch. In this section we consider the other interference, the one that arises from the contributions of the N periods that make up a given grating. There was a passing remark to this earlier on where it was stated that the effect of N periods is that the yield from one period has to be multiplied by N^2 in order to obtain the total yield from the grating. This is true, provided that the contributions of the N periods are all in phase. This condition is satisfied when the observer is at 'infinity' or sufficiently far away from the grating so that the latter can be considered as a point source. In practice, of course, this is almost never satisfied but we need to consider what might be deemed to be 'infinity' and what the consequences are when a measurement is made at a position that is not far away from the grating surface.

It is helpful to start with a purely qualitative argument. The interaction of a given beam with a given grating will release in the 2π solid angle above the grating a certain amount of energy in the form of SP radiation; it is assumed that the beam does not impinge on the grating. What is the wavelength distribution of the energy measured by a detector? Consider the following 'thought' experiment whereby a detector with a fixed solid angle is first placed rather close to the grating but at a fixed angular position; the measured energy will be spread over a rather broad wavelength band. As the detector is moved further away from the grating, while maintaining the same angular position, the accepted wavelength band will become narrower and narrower until at 'infinity' the detected radiation will be essentially monochromatic, as described by the SP condition. Another way of expressing this reasoning would be to say that if the detector is not at infinity, the measured energy at the relevant SP wavelength will be lower than that expected from theory because a certain fraction of it will have spread out to adjacent wavelengths. It is possible to calculate this loss and a sample calculation is shown in this section; it can obviously be adapted to suit the specific conditions of any optical system.

An SP grating can be seen as the equivalent to an array of independent oscillators, whose total number is equal to the number of periods on the grating and which are excited sequentially by the passing electron (or electron bunch). One period of the grating constitutes a single oscillator. There is thus a timing condition that has to be satisfied for the excitation of these oscillators. It is assumed, that the observation is made at a point P, which is located at a distance (r) from the centre (O) of the grating. The derivation of analytic expressions for the interference effects is unnecessary since the main points of the discussion can be clarified by a simple geometric picture, such as the one shown in fig. 2.3, which is a schematic of *one-half* of a grating having a period (l).

There are N periods in this *half* of the grating, hence its length is $L = Nl$; the figure is not drawn to scale. The symbols O, Q, R, etc. represent successive periods of the grating and are spaced (l) apart. The distances d_1, d_2, etc. are the additional path lengths to point P compared to the path length OP. Hence, they introduce a phase difference determined by the time taken for an electron travelling left to right to cover the distance QO, minus the time taken by the light to travel the distance d_1:

$$\frac{l}{\beta c} - \frac{d_1}{c} = \frac{l}{\beta c} - \frac{l cos\theta}{c}$$

If this time difference is equal to an integer multiple of the period (T) of the wave, i.e. if:

$$\frac{l}{\beta c} - \frac{l cos\theta}{c} = nT \tag{2.4}$$

then their phase difference is an integer multiple of 2π and there is constructive interference between rays emanating from points O and Q. This will happen when the detection point P is far away from the grating, at 'infinity', and eq. (2.4) will be satisfied for a given period (wavelength); this is, of course, the basic SP relationship between wavelength and emission angle.

In the schematic of fig. 2.3, where the detector is not at infinity, there is a wavelength-dependent phase difference between the rays emanating from O, Q, R, etc. and each period contributes, with the appropriate phase, to the total field at P. The resultant intensity is proportional to the square of the complex series, which

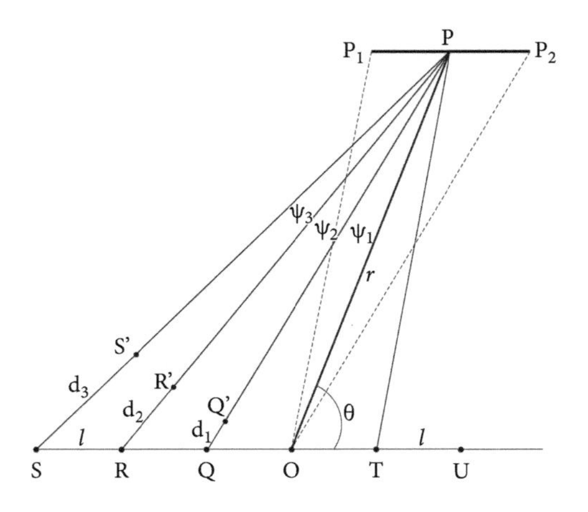

Fig. 2.3 Schematic of the path differences between grating and detector aperture.

can be calculated by noting that the path difference d_1 is:

$$d_1 = l\frac{\cos\left(\theta \mp \frac{\psi_1}{2}\right)}{\cos\left(\frac{\psi_1}{2}\right)}$$

with:

$$\psi_1 = asin\left[\frac{sin\theta}{\sqrt{1 + \frac{r^2}{l^2} \pm \frac{2r}{l}cos\theta}}\right] = atan\left[\frac{lsin\theta}{r \pm lcos\theta}\right]$$

and similarly, for d_2, d_3 etc. [If $\theta > 90^0$, i.e. P is on the LH side of the centre-line of fig. 2.3, then the lower of the +/− signs should be used in the above expression. A similar change of signs, depending on the position of P relative to the centre-line, is required when dealing with the RH side of the grating.] The maximal value ($\Delta\alpha$) of the phase difference is that between the centre (O) and the first period on the left edge of the grating:

$$\Delta\alpha = 2\pi\frac{(d_N - Nl\,cos\theta)}{\lambda} = 2\pi L\frac{sin\theta\,tan\frac{\psi_N}{2}}{\lambda}$$

This sum can then be compared to the value that it would be expected to have if point P were indeed at infinity. It is assumed implicitly that there are no significant contributions arising from path differences in the azimuthal direction. As r tends to infinity, $\psi \rightarrow 0$, the lines to the observation point are parallel to each other, $d_2 = 2d_1$ etc. and there is strict proportionality between the d_i values; there is, thus, no phase difference between the oscillators and the above expressions become the SP formula.

The actual experimental situation is a bit more complicated because detection is not at a single point P but along the line P_1P_2 in fig. 2.3, which can represent the effective detector aperture. The interference effects at each of these angular positions around θ can be estimated by replacing the $lcos\,(\theta)$ term in the above expressions with $lcos\,(\theta + d\theta)$ and re-evaluating the complex sum. For the purposes of the present numerical example, we use the parameters of the optical system of [3] where the angular acceptance was $\theta \pm 6°$ and the detector was at a distance of r = 303 mm from the grating; therefore, at $\theta = 90°$ the accepted range is 84–96°. The significance of the interference effects can be expressed as the ratio of two intensities: the first is the one expected with the detector placed at a distance r = 303 mm from the grating and the second with the same detector at 'infinity', which is assumed to occur at a distance r = 50 m. The results are shown in fig. 2.4 as a function of wavelength for four gratings of different periodicity. Evidently, the shorter the wavelength the more important it becomes to consider the actual position of the detector and to make any necessary corrections to the measured signal.

Therefore, from the experimentalists' point of view it is important to know whether their detector is in the far field region, i.e. at infinity, or whether they need

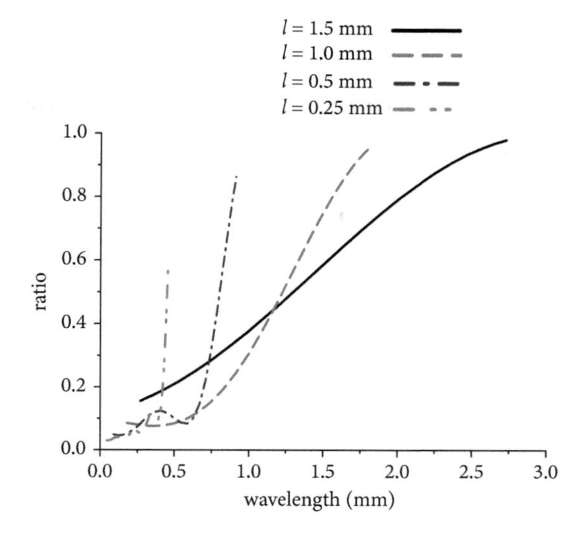

Fig. 2.4 The ratio of the signal intensity expected with the detector placed at a distance of 303 mm, relative over that expected with the detector at 'infinity'. The ratios are plotted as a function of the wavelength of the emitted radiation, for four gratings of different periodicity.

to carry out the type of calculation outlined above. According to [4] the far field criterion is satisfied if the detector is placed at a distance (L_d) such that:

$$L_d \gg LNn\,(1 + cos\theta) \tag{2.5}$$

where L is the grating length, N the number of periods, n is the order of emission, and θ the observation angle. The curves of fig. 2.3 were calculated for gratings of length 36 mm. If we were to consider one specific wavelength, say $\lambda = 1.5$mm, then for the $l = 1.5$ mm grating the detector would have to be placed at $\theta = 90^0$ and according to (2.5) infinity would be at distances much greater than 864 mm, while for the $l = 1.0$ mm grating the detector would have to be placed at $\theta = 120^0$ but the requirement of (2.5) would be satisfied at distances greater than 648 mm. Thus, from the perspective of monochromaticity *alone*, the use of the $l = 1.0$ mm grating is preferable, as seen also in the plots of fig. 2.4. There are, of course, other considerations that may argue in favour of the longer period grating.

In summary, and within the accuracy of this calculation, interference effects in the optical system used in ref. [3] have a small influence on the measured intensity for wavelengths in the few mm range but become progressively more severe at shorter wavelengths; at a wavelength of about 250 μm, less than 10% of the energy calculated on the basis of the far field assumption would have been actually observable at that specific wavelength. The rest would have been smeared out and distributed over the wavelengths that fall within the $\theta \pm 6^0$ band. Ultimately, the details of the optical system used in a specific experiment will determine the validity of the 'observer at infinity' assumption and the requirement for any corrections.

2.3 Higher-order emission

Before discussing higher-order emission, it is helpful to rewrite the basic relationship between wavelength and emission angle:

$$l\left(\frac{1}{\beta} - \cos\theta\right) = \lambda n \tag{2.6}$$

and to draw the readers' attention to the (unsurprising) similarity of this expression with that of the diffraction from a grating [1]. Thus, although SP radiation is often discussed as a potential source of monochromatic radiation, this is obviously not strictly true; at any observation angle θ, there are many orders of emitted radiation, each with its own intensity. These can be separated by suitable filtering but there are some interesting questions that come to mind: (a) are there circumstances under which a desired wavelength may be more intense as, say, a second harmonic at an angle θ_2, rather than as the fundamental at θ_1? (b) if so, would there be an advantage in opting for 2nd order emission? (c) for a given observation angle, what are the comparative intensities of the fundamental $n = 1$ and of a higher order, say $n = 2$?

A semi-analytical discussion of the first question would be as follows. To begin with, it is easy to confirm by manipulation of (2.6) that if a certain wavelength can be observed as 1st order emission at θ_1 and as nth order emission at θ_n, the relationship between these angles is given by:

$$1 - \beta cos\theta_n = n\,(1 - \beta cos\theta_1)$$

This relationship is plotted in fig. 2.5 for orders 2 to 5 and for $\beta \cong 1$; also shown, with the dashed line, is the case of a low energy (50 keV) electron beam where $n = 2$ and $\beta = 0.416$.

Thus, for a highly relativistic beam ($\beta \cong 1$) the same wavelength can be observed as 1st order emission at 60^0 or as 2nd order emission at 90^0 or as 3rd order emission at 120^0 and so on. However, for a low energy beam ($\beta = 0.416$) 1st order emission at 30^0 would correspond to 2nd order at 132^0 etc. It is worth noting that for a relativistic beam all orders tend to pile up in the forward direction; in fig. 2.5, the same wavelength could be observed as fundamental or higher order emission in the narrow angular range of approximately $5–11^0$.

Equation (1.7a), which gives the single-electron yield for the case where the 'grating of infinite width' approximation is valid, can be rewritten as:

$$\frac{dI}{d\Omega} = \frac{2\pi e^2 \beta^3 Z n^2}{l^2(1 - \beta cos\theta)^3} e^{-\frac{2x_0}{\lambda_e}} R^2 \tag{2.7}$$

Taking into account the previously stated relationship between θ_1 and θ_2, it is then possible to derive an expression for the ratio (κ) of the radiated intensities at *the*

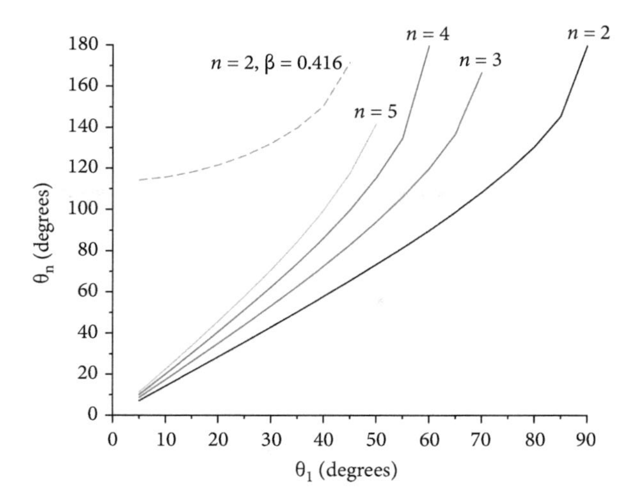

Fig. 2.5 A given wavelength can be observed as 1st order emission at an angle θ_1 or as n^{th} order emission at another angle θ_n. The solid lines are for a relativistic beam ($\beta \cong 1$) and the dashed one for a low energy beam ($\beta = 0.416$).

same wavelength λ but at angles θ_1 and θ_2 corresponding, say, to $n = 1$ and $n = 2$:

$$\kappa = \frac{1}{2} e^{-\frac{4\pi x_0 sin\phi(sin\theta_2 - sin\theta_1)}{\lambda}} \frac{R_2^2}{R_1^2} \tag{2.8}$$

The above expression assumes that the azimuthal angle ϕ is not zero (this is always true for a detector of finite entrance aperture) and that $\beta \cong 1$, so that the evanescent wavelength $\lambda_e \cong \frac{\lambda}{2\pi sin\theta sin\phi}$, i.e. that the beam is relativistic. R_1^2 and R_2^2 are the efficiency factors at the corresponding observation angles. Analogous expressions can be derived for the ratio of intensities at 1st and nth order emission.

The question then is whether κ is smaller or greater than unity? Since $\theta_3 > \theta_2 > \theta_1$, κ would always be less than 1, unless $\frac{R_2^2}{R_1^2} > 1$. The grating efficiency factors R^2 are a function of angle but unfortunately, as stated in Chapter 1, there are no analytic expressions available and hence, it is not possible to give an answer of general validity. Therefore, it is best to carry out a numerical study assuming, for example, a single electron having $\gamma = 100$ and travelling at a height of 0.5 mm above an echelette grating having a period of 1.0 mm and a blaze angle of 30^0. The results of the simulation are shown in figs. 2.6–2.8 which also help to summarize the discussion of this section. The evaluation of the grating efficiency factor R^2 as a function of the observation angle (always relative to the beam direction) shows a minimum at twice the blaze angle, i.e. at 60^0 (fig. 2.6).

This is in complete analogy with the null observed at the angle of specular reflexion in the case of diffraction radiation [6]. The normalized single electron yield as a function of wavelength is shown in fig. 2.7, which also answers the question

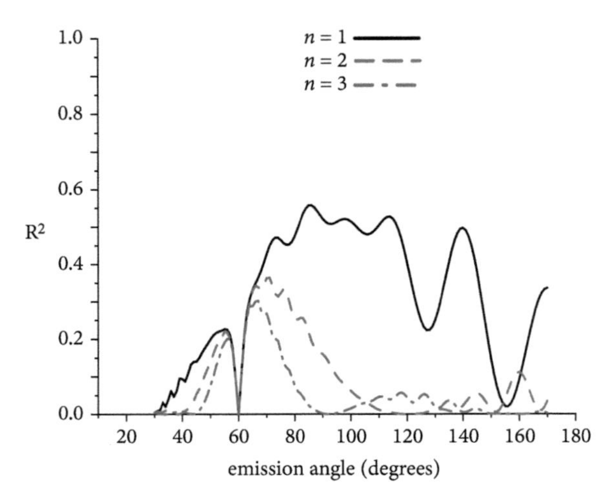

Fig. 2.6 The grating efficiency factor R^2 as a function of emission angle for orders 1, 2, and 3. The simulation has been carried out for an echelette grating of infinite width with a blaze angle of 30^0 and a period of 1.0 mm.

posed earlier on, whether it is possible to obtain (for a given wavelength) higher yield from, say, 2nd or 3rd order emission.

The answer is 'yes' and this can be understood in simple terms as occurring when the grating efficiency happens to be significantly higher at the emission angle of the 2nd (or 3rd) order. In the example of fig. 2.7 and for a wavelength of 250 μm, there is nearly 50% more energy at $n = 2$ compared with $n = 1$. In this specific example, 1st order emission for $\lambda = 250$ μm would occur at $\theta \cong 41^0$ while the 2nd order would appear at $\theta \cong 59^0$. Moreover, utilization of higher-order emission, whenever the calculations indicate that it might be of comparable intensity to that of

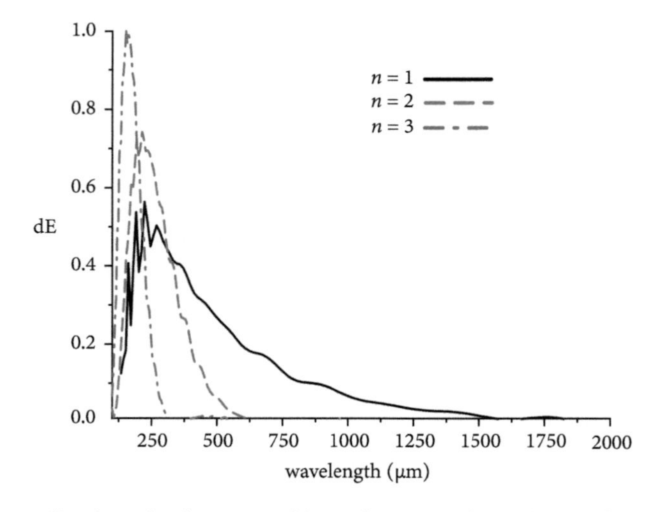

Fig. 2.7 The normalized single electron yield as a function of wavelength for orders 1, 2, and 3; the electron impact parameter is $x_o = 0.5$ mm and the other grating parameters are as in fig. 2.4.

the 1st order, results in an improvement in the monochromaticity of the radiation since:

$$\frac{\Delta\lambda}{\lambda} = \frac{1}{nN}$$

where N is the number of periods in the grating. This is the familiar expression for the resolving power of a grating (see ref. 5 for a lucid and concise explanation). In the case of SP radiation, the expected improvement in wavelength resolution has been observed experimentally [7].

The preceding discussion was centred on the comparison of the intensities of a given wavelength, observed at two different emission angles. A related question would be whether at a given *observation angle*, the 1st order emission is indeed the strongest or whether, say, the 2nd order which would now correspond to a different wavelength, might be more intense. The answer to this is clearly affirmative, as shown in the numerical simulation of fig. 2.8.

However, there is a word of warning: all the above simulations are based on the 'single-electron' yield and are valid for DC beams or beams that are sufficiently dilute so that coherence effects are negligible. The situation will be very different for tightly bunched beams where coherence will start playing a dominant role and the yield curves of fig. 2.8 will be altered drastically, as discussed in the previous section; the black line of fig. 2.8, for example, corresponds to longer wavelengths and will be boosted by coherence to a much higher extent than the other two. Higher-order emission will be discussed again in Chapter 6, in the context of SP radiation as a source of tuneable far-infrared (FIR) radiation.

In summary, although there is no general rule about the intensity of high-order emission and although most of the published work has been carried out with 1st

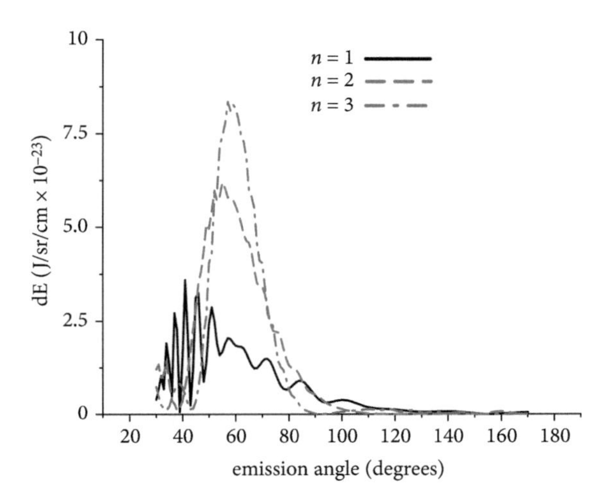

Fig. 2.8 The single-electron yield as a function of emission angle for orders 1, 2, and 3; the electron impact parameter is $x_o = 0.5$ mm and the other grating parameters are as in fig. 2.6.

order emission, it is worth carrying out the necessary calculations in order to explore the possibilities of higher-order emission.

Before concluding the theoretical discussion of the SP radiation, it is interesting to note the existence of the inverse effect, namely the 'inverse SP radiation' or ISP. The reasoning is as follows: at a descriptive and non-mathematical level, the SP process can be seen as the interaction of a particle beam with an electromagnetic wave, mediated by the presence of the grating; energy is extracted from the particle beam and emitted as radiation. However, it also possible to have the inverse process whereby energy from an *incident* electromagnetic wave is transferred to the particle beam, again mediated by the grating, resulting in the acceleration or deceleration or just bunching of the particles. In the 1980s there was considerable interest in the ISP effect, primarily from Japanese research groups which have studied ISP both theoretically and experimentally, with the aim of providing another method for accelerating charged particles [8, 9]. It is probably easier to visualize the ISP process as a situation where the incident laser light produces evanescent modes along the grating surface and if one of these modes has a phase velocity matching the beam velocity, then it is possible to transfer energy to the particle beam. However, these 'slow' waves are evanescent perpendicularly to the grating surface and it is necessary to bring the beam close to the surface, just as in the case of the SP emission. The evanescent wavelength λ_e, introduced in section 1.1a, is equally relevant in ISP as in SP and it indicates that the use of laser light with longer wavelengths, say in the sub-millimetre range, would be an advantage.

Having established the essential elements of the SP radiation theory, it is now possible to proceed to a detailed discussion of the applications of this radiative process. These can be divided into two categories: in the first category, the radiation is used as a probe to investigate some property of the electron beam that gave rise to this radiation. This is a rather subtle physical problem which merits a more detailed discussion because of its importance in the design of high energy particle accelerators. This is the subject matter of the following two chapters. Alternatively, the SP process could be used as a tuneable source of radiation for investigations in another branch of physics, or science in general; this is probably the most obvious potential use of SPr and will be discussed in Chapters 5 and 6.

References

1. R. Feynman, R. Leighton, and M. Sands, 'Lectures on Physics', Vol. **I**-30, Addison Wesley (1966)
2. J.S. Nodvick and D.S. Saxon, Phys. Rev. **96** (1954) 180
3. H.L. Andrews et al, Phys. Rev. Accel. & Beams **17** (2014) 052802
4. D.V. Karlovets and A.P. Potylitsyn, JETP Letters **84** (2006) 489
5. R.P. Feynman, R.B. Leighton and M. Sands 'The Feynman Lectures on Physics', Addison Wesley Vol. **I**-30 (1966), section 30–3

6. P. Henri et al, Phys. Rev. E **60** (1999) 6214
7. A. Aryshev et al, Phys. Rev. Accel. & Beams **20** (2017) 024701
8. K. Mizuno, J. Pae, T. Nozokido, and K. Furuya, Nature **328**(6125) (1987) 45
9. J. Bae et al, Japanese Journal of Applied Physics **27**(3) (1988) 408

3
Electron Beam Diagnostics

In the great majority of particle accelerators, the beam is delivered in the form of bunches. These can either be equally spaced in time or they may be grouped together in bunch 'trains', where a burst of bunches is separated from the next train by a fixed time period. Both from the beam user's and the accelerator designer's point of view, it is important to know the shape of the bunch and to be able to control it. Discussion of the issues associated with the transverse dimensions of the bunch (focussing, size measurement, etc.) is not within the scope of this work. Measurements in the third (time) dimension of the bunch, however, are closely related to a radiative process such as SP radiation and are the subject of this chapter.

3.1 The time profile of ultra-short bunches

The time duration (length) of each bunch varies according to the accelerator type and the method used to produce the bunch but, as a general comment, it is true to say that the tendency is towards high charge density and hence, ever shorter bunches, in the ps and sub-ps region. In many cases it is important to know not only the bunch length (the accelerator designers already have some idea about this) but also the details of the longitudinal, i.e. the time, profile of the bunch. In the case of linear colliders, for example, the longitudinal profile of the colliding bunches determines the generation of beamstrahlung at the collision point and affects the overall background [1]. Accelerator-based light sources such as X-ray Free Electron Lasers (FELs) also require high current bunches, which means bunch lengths in the few fs range. At the same time, the recent progress in plasma wakefield acceleration holds the promise of very short bunches and very high accelerating fields. In order to be able to understand the physics of the generation of the bunch in the plasma or the evolution of the bunch during compression and transport, it is essential to be able to determine its time profile accurately. There are a number of methods that are currently available, but all of them have advantages and disadvantages and they all become progressively more difficult in the fs regime. A recent, comprehensive review of beam diagnostics for plasma-based accelerators can be found in [2].

The oldest and probably most straightforward method is the use of a streak camera. These are commercially available electro-optic devices where the electron bunch is used to produce in the first instance a light pulse by some radiative process such as optical transition radiation (OTR) or Cherenkov radiation (CR). The light pulse then impinges on a photocathode and the emitted photoelectrons are

Smith-Purcell Radiation. George Doucas, Oxford University Press. © George Doucas (2025).
DOI: 10.1093/9780198951360.003.0004

transversely deflected by a rapidly varying electric field; the time information is thus transformed into a spatial profile which is much easier to analyse by means, for example, of a micro-channel plate (MCP). These are devices for measuring the bunch length rather than details of the time profile. Time resolutions below 1 ps can be achieved [3–5]. The same idea of transforming the time information into a different domain is used in the transverse deflecting cavities. The bunch is deflected by a rapidly varying field into one of the transverse dimensions. One might visualize this as the bunch being 'pitched' or 'yawed' and at the same time deflected out of the beam path, so that it can be viewed on a suitable screen downstream. This method can provide very accurate (down to a few fs) information about the time profile, limited only by the dimensions of the bunch in the transverse direction into which it is deflected. However, it does require significant space for the installation of the cavity and significant power for its operation [6, 7].

Another possibility is to map the time information onto energy. This is applicable to multi-section linear accelerators where the last section can be run in such a fashion that the field is varying rapidly as the bunch passes through and the particles can thus receive an energy boost proportional to their time of arrival. The time profile of the bunch is thus transformed into an energy profile, which can be sampled by an energy spectrometer. The limit is the inherent energy spread of the beam and the speed of ramping up the field [8–10]. Time resolutions of the order of 100 fs have been achieved. A further alternative is to change the time information into a change of polarization of a laser pulse. This is the Electro-Optic Sampling (EOS) method. The basic physical idea is as follows: the electric field associated with a relativistic electron bunch is almost entirely transverse and can be used to induce transient birefringence in a crystal such as GaP or ZnTe; this is the electro-optic effect. The particle beam passes close to the crystal but does not hit it and is not affected by this interaction. If, at the same time, a femtosecond laser beam of known polarization (the 'probe' pulse) passes through the crystal, its polarization will be altered by the induced anisotropy in the crystal. This change in polarization, which depends on the electric field of the bunch (i.e. its time profile), can be detected as a function of probe delay and thereby used to determine the profile of the bunch. There are several variants of EOS and the best resolution is better than 32 fs [11–15]. EOS is a powerful diagnostic method but the installation is cumbersome and the issues of timing are critical.

Apart from the above methods, it is possible to determine the temporal profile of a bunch by measurements carried out in the frequency domain, namely by spectral analysis of the radiation emitted by the bunch. SP radiation is one such radiative process but it is not the only one; transition, diffraction, and synchrotron radiation could also be used for this purpose. The way this can be achieved will be described in the rest of this chapter. It is interesting to note that the previously mentioned diagnostic methods are easy to understand physically (the time profile is 'mapped' on another, easier to measure, physical quantity) but are complicated in their implementation. The situation is almost the opposite for measurements in the frequency domain: the implementation is relatively simple but the physics is rather opaque. The steps by which one can get from the spectral analysis of the emitted radiation to the profile of the bunch that gave rise to this radiation will be described in the next section. Since the bunch profile is of importance in accelerator physics, the subject will be discussed in some detail.

3.2 Reconstruction of the time profile by measurements in the frequency domain

Before proceeding with the details of the calculation, it is useful to outline the problem.

a. Measurements in the frequency domain will only give the magnitude of the Fourier transform (FT) of the time profile, *but not* the phase.
b. It is possible to recover some information about the missing phase by means of the Kramers–Kronig (KK) relations. These are integral formulae that relate the real and imaginary parts of a linear, causal response function [16] and can be used to estimate the minimal, or canonical, phase (θ). They are also known as dispersion relations and are used widely in many branches of physics, one notable example being the study of the optical properties of solids, where measurements of the reflectance can be used to determine the absorption coefficients. The 'response' function can be understood as a function that describes the response of a system at a time t, following an excitation by a delta function at time $t = 0$. 'Linear' means that the output is a linear function of the input and 'causal' means that there can be no output before the input. The equivalence of causality and the dispersion relations has been discussed in detail in the work of Toll [17].
c. In order to apply KK, it is necessary to know the magnitude (ρ) of the FT over all frequencies; since this is not possible, it is desirable to measure ρ at as many frequency points as possible and then interpolate between these points and extrapolate to zero frequency and to infinity.
d. It is possible that there may be additional contributions to the minimal phase. These are known as Blaschke phases and they must be added to the minimal phase. However, their existence (or otherwise) cannot be known *a priori*; thus, the minimal phase may or may not be the true phase.
e. Therefore, frequency domain measurements will give a good approximation to the time profile but cannot be guaranteed to give the exact profile.

In spite of these limitations, frequency domain methods are used widely in time profile reconstruction because of their relative simplicity. The first question that arises is how these ideas can be applied to the case of the bunch temporal profile.

3.3 The minimal phase

The starting point of the calculation of the minimal phase is the expression for the radiative yield $\left(\frac{dI}{d\Omega}\right)_{Ne}$ from a bunch of N_e electrons. This was derived in Chapter 2 but is given again here for convenience.

$$\left(\frac{dI}{d\Omega}\right)_{N_e} = \left(\frac{dI}{d\Omega}\right)_1 \left(N_e S_{inc} + N_e^2 S_{coh}\right) \tag{3.1}$$

where $\left(\frac{dI}{d\Omega}\right)_1$ is the single electron yield and S_{coh} is the coherent integral:

$$S_{coh} = \left|\frac{1}{\sigma_x\sqrt{2\pi}}\int_0^{\infty} e^{\frac{-x}{\lambda_e}}\, e^{\frac{(-x-x_0)^2}{2\sigma_x^2}}\, dx\right|^2 \left|\frac{1}{\sigma_y\sqrt{2\pi}}\int_{-\infty}^{\infty} e^{-ik_y y}\, e^{-\frac{(y-y_0)^2}{2\sigma_y^2}}\, dy\right|^2 \left|\int_0^{\infty} e^{-i\omega t}\, T(t)\, dt\right|^2$$

We assume that we are operating in the coherent regime, i.e. in the wavelength region where the wavelength of the radiation is approximately equal or greater than the bunch length. In this case, the term $N_e\, S_{inc}$ is negligible compared with $N_e{}^2 S_{coh}$ and can be neglected. Moreover, it is usually possible to focus the beam down so that its transverse dimensions are smaller than the wavelengths of interest, in which case the first two integrals of S_{coh} are close to unity. In other words, it is reasonable to assume transverse coherence. Therefore,

$$S_{coh} \cong \left|\int_0^{\infty} e^{-i\omega t}\, T(t)\, dt\right|^2$$

$$\left(\frac{dI}{d\Omega}\right)_{N_e} \cong \left(\frac{dI}{d\Omega}\right)_1 \left(N_e^2 S_{coh}\right)$$

Assuming that the single electron yield is known (this is the task of any theoretical description of this radiative process) and that the number of electrons in the bunch is also known, then a measurement of the radiated power gives ρ^2, the square of the magnitude of the FT ($\tilde{T}(\omega)$) of the time profile of the bunch:

$$\rho^2 = \left|\tilde{T}(\omega)\right|^2$$

The magnitude (ρ) of the FT of the time profile was introduced in Chapter 2 and is sometimes referred to as the longitudinal form factor.

Since there is no radiation emitted before the arrival of the bunch, causality is satisfied. The time profile is of finite duration and can always be shifted along the time axis so that $T(t) = 0$ for $t < 0$. Since the radiated energy is proportional to the square of the electron's field, the field due to the N_e electrons in the bunch in the coherent part of the spectrum is related, according to (3.1), to the single electron field $\bar{E}_1$ by:

$$\bar{E}_{Ne}(\omega) = N_e\bar{E}_1(\omega)\int_0^{\infty} e^{-i\omega t}\,T(t)\,dt = N_e\tilde{T}(\omega)\,\bar{E}_1(\omega) \tag{3.2}$$

The role of the transverse size of the bunch has been neglected, on the assumption that the transverse size is sufficiently small, as mentioned earlier. One could then argue [18] that this is a situation where the KK relations are indeed applicable: the stimulus, or excitation, is $\bar{E}_1$ and the response is $\bar{E}_{Ne}$ while the response function is $\tilde{T}(\omega)$, the FT of the longitudinal profile. The quantity $\omega = 2\pi\nu$ is the (angular) frequency of the radiation.

The analysis that follows is based on the work of Wooten [19] and Grimm and Schmüser [20]. The determination of the minimal phase requires the continuation of $\tilde{T}(\omega)$ in the complex frequency plane $\hat{\omega} = (\omega_r, \omega_i)$ and therefore $\tilde{T}(\omega)$ becomes $\tilde{T}(\hat{\omega})$. This function is analytic and bounded in the lower half of the complex plane ($\omega_i < 0$) because the term $e^{-i\hat{\omega}t}$ is analytic and bounded for $\omega_i < 0$. (N.B. If the FT had been defined as $e^{i\hat{\omega}t}$, then one would have to use the upper half plane). In order to introduce the phase angle (θ) we write:

$$\tilde{T}(\hat{\omega}) = \rho(\hat{\omega})\,e^{i\theta(\hat{\omega})}$$

and

$$ln\tilde{T}(\hat{\omega}) = ln\rho(\hat{\omega}) + i\theta(\hat{\omega})$$

where ρ and θ are real. Since $\tilde{T}(\hat{\omega})$ is analytic, $ln\tilde{T}(\hat{\omega})$ will also be analytic, provided that $ln\rho(\hat{\omega})$ does not go to zero. This, however, cannot be excluded in the case of the bunch profile reconstruction: the form factor must vanish at high frequencies because the bunch profile must only have features of finite width. Therefore, it is necessary to use another function for the integration, such as the one below:

$$f(\hat{\omega}) = \frac{(\omega_0\hat{\omega} + 1)}{(\hat{\omega}^2 + 1)(\omega_0 - \hat{\omega})}ln\tilde{T}(\hat{\omega})$$

The function $f(\hat{\omega})$ is analytic because it is the product of two analytic functions, namely $ln\tilde{T}(\hat{\omega})$ and a fraction that can also be shown to be analytic. It has one pole in the lower half plane, at $\hat{\omega} = -i$, and another one on the boundary of the integration contour, at $\omega = \omega_0$. According to the residue theorem, the value of the contour integral $\oint f(\hat{\omega})\,d\hat{\omega}$ over the contour shown in fig. 3.1 is equal to:

$$\oint f(\hat{\omega})\,d\hat{\omega} = 2\pi i.Residue(-i) = -i\pi ln\tilde{T}(-i)$$

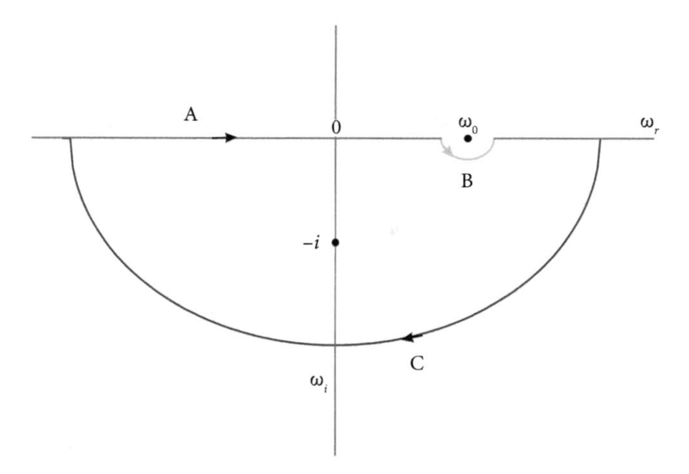

Fig. 3.1 Contour of integration over the lower half of the complex frequency plane.

Note that since by definition.

$$\tilde{T}(\omega) \approx \int_0^{\infty} e^{-i\omega t}\, T(t)\, dt$$

$\tilde{T}(-\omega) = \tilde{T}^*(\omega)$ and $\rho(\omega) = \rho(-\omega)$; also, that $\tilde{T}(-i)$ is a real number.

The contour of integration is now split into three parts, as shown in fig. 3.1. In part A, along the real axis, the integral is:

$$P\int_{-\infty}^{+\infty} f(\omega)\, d\omega = P\int_{-\infty}^{+\infty} \frac{(\omega_0\omega + 1)\, ln\rho(\omega)}{(\omega^2+1)(\omega_0-\omega)} d\omega$$

The symbol P signifies the principal value of the integral, since we have excluded from the integration the singularity at $\omega = \omega_0$. The frequency is real along this sector of the integration contour, hence $\hat{\omega}$ has been replaced with the (real) ω. Part B of the contour integral, the small semicircle around the singularity $\omega = \omega_0$, can be evaluated by setting $\hat{\omega} = re^{i\psi}$ and taking the limit as $r \to 0$. Alternatively, one could argue that if a single pole is inside the boundary it contributes $2\pi i.Residue$ to the value of the integral whereas if it is outside the contour it contributes nothing; therefore, if it is *on* the boundary it will contribute $\pi i.Residue$. In our case this contribution amounts to:

$$i\pi ln\tilde{T}(\omega_0)$$

This leaves part C of the contour integral, around the large semicircle. As mentioned earlier, $\tilde{T}(\hat{\omega}) \to 0$ when $\omega_r \to \infty$ because the bunch cannot contain infinitely fine structures, and also when $\omega_i \to -\infty$ because $\tilde{T}(\hat{\omega})$ is bounded in the lower half of the plane. Therefore, $\tilde{T}(\hat{\omega}) \to 0$ when $|\hat{\omega}| \to \infty$ and it can be written as $\tilde{T}(\hat{\omega}) \leq a\hat{\omega}^{-b}$, where a and b are real positive constants. Then the absolute value of the integrand becomes:

$$|f(\hat{\omega})| = \left|\frac{(\omega_0\hat{\omega} + 1)(lna - bln\hat{\omega})}{(\hat{\omega}^2 + 1)(\omega_0 - \hat{\omega})}\right| \to 0 \text{ when } |\hat{\omega}| \to \infty$$

as can be verified by application of L'Hôpital's rule, and the integral along part C tends to zero. The overall value of the contour integral is the sum of parts A, B, and C, and hence:

$$P\int_{-\infty}^{+\infty} \frac{(\omega_0\omega + 1)}{(\omega^2 + 1)(\omega_0 - \omega)} ln\rho(\omega) + i\pi ln\tilde{T}(\omega_0) = -i\pi ln\tilde{T}(-i)$$

Since $ln\tilde{T}(\omega_0) = ln\rho(\omega_0) + i\theta(\omega_0)$, we take the real part of the above equation to obtain the relation between phase and magnitude:

$$\theta(\omega_0) = \frac{1}{\pi} P \int_{-\infty}^{+\infty} \frac{(\omega_0\omega + 1)\, ln\rho(\omega)}{(\omega^2 + 1)(\omega_0 - \omega)} d\omega \tag{3.3}$$

It is possible to restrict the integration of (3.3) to positive values of frequency by using the fact that $\tilde{T}(-\omega) = \tilde{T}^*(\omega)$ and $\rho(\omega) = \rho(-\omega)$. Hence, by splitting the domain of integration into $(-\infty, 0)$ and $(0,\infty)$ it is easy to confirm that:

$$\theta(\omega_0) = \frac{2\omega_0}{\pi} P \int_0^{+\infty} \frac{ln\rho(\omega)}{\left(\omega_0^2 - \omega^2\right)} d\omega \tag{3.4}$$

There is a singularity at $\omega = \omega_0$ which can be removed by subtracting from (3.4) the quantity:

$$\frac{2\omega_0}{\pi} P \int_0^{+\infty} \frac{ln\rho(\omega_0)}{\left(\omega_0^2 - \omega^2\right)} d\omega = 0$$

That the above expression tends to zero can be verified by splitting the domain of integration into $(0,\ \omega_0 - \delta)$ and $(\omega_0 + \delta, \infty)$ and taking the limit as $\delta \to 0$. Therefore, the final expression for the minimal phase at an angular frequency ω_0 becomes:

$$\theta(\omega_0) = \frac{2\omega_0}{\pi} \int_0^{+\infty} \frac{ln\left[\rho(\omega)/\rho(\omega_0)\right]}{\left(\omega_0^2 - \omega^2\right)} d\omega \tag{3.5}$$

There is now no singularity at $\omega = \omega_0$, as can be verified by L'Hopital's rule.

If, as stated earlier, $\tilde{T}(\omega)$ is the FT of the time profile $T(t)$, then:

$$T(t) = \frac{1}{2\pi}\int_{-\infty}^{\infty} \tilde{T}(\omega)\, e^{i\omega t}\, d\omega = \frac{1}{2\pi}\int_{-\infty}^{\infty} \rho(\omega)\, e^{i\omega t} d\omega$$

and since $\rho(\omega) = \rho(-\omega)$, the integration can be restricted to positive frequencies. Hence, the time profile can be recovered from either of the two equivalent expressions below, the first in terms of angular frequency (ω) and the other in terms of frequency (ν):

$$T(t) = \frac{1}{\pi}\int_{-\infty}^{\infty} \rho(\omega) \cos[\omega t + \theta(\omega)]\, d\omega \tag{3.6}$$

or:

$$T(t) = 2\int_{0}^{\infty} \rho(\nu) \cos[2\pi\nu t + \theta(\nu)]\, d\nu \tag{3.6a}$$

The minimal phase is a function of frequency ω, but note that a simple linear relationship of the type $\theta = a\omega$, where a is a constant, does not provide any information about the asymmetry of the profile; it is simply a translation in time of the whole profile, as can be verified easily from eq. (3.6). It is the non-linear relationship between θ and ω that determines bunch asymmetry.

3.4 Limitations of the KK reconstruction

There are limitations in the KK reconstruction of the longitudinal shape of a bunch. The first is that the procedure cannot distinguish between two bunches, each of which is a time reversal of the other (see fig. 3.2).

Let $T_1(t), T_2(t)$ be the two time-reversed profiles that extend from $t = 0$ to $t = t_f$ and $\tilde{T}_1(\omega), \tilde{T}_2(\omega)$ the corresponding FTs. Because of their time reversal relationship, $T_1(t) = T_2(t_f - t)$ and $T_2(t) = T_1(t_f - t)$. Then, by changing $t \rightarrow t_f - t$:

$$\tilde{T}_2(\omega) = \int_{-\infty}^{\infty} T_2(t)\, e^{-i\omega t} dt = \int_{-\infty}^{\infty} T_2(t_f - t)\, e^{i\omega t} e^{-i\omega t_f} dt$$

$$= e^{-i\omega t_f}\int_{-\infty}^{\infty} T_1(t)\, e^{i\omega t} dt = \tilde{T}_1^*(\omega)\, e^{-i\omega t_f}$$

and hence:

$$|\tilde{T}_2(\omega)| = |\tilde{T}_1^*(\omega)| = |\tilde{T}_1(\omega)|$$

Therefore, the two profiles have FTs of the same magnitude and cannot be separated by the analysis presented here.

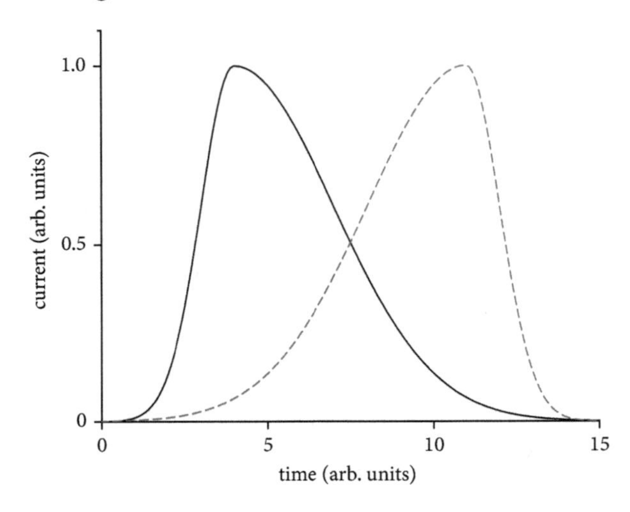

Fig. 3.2 Schematic of two bunches, each of which is a time reversal of the other.

Much more important, however, is the question of any possible zeros in the lower half of the complex plane and their effect on the phase. The main application of the KK analysis was in the study of the complex reflectivity of solids where the possibility of zeros in ρ can be excluded. However, this is not always true in the case of the bunch profile problem, where zero values of ρ can exist. A detailed discussion of this subject is beyond the scope of this work since, as mentioned earlier, the existence of zeros cannot be known *a priori* and therefore there is very little that an experimentalist can do about this. It is adequate for our purposes to point out a couple of basic facts about these zeros:

a. Because of the fact that $\tilde{T}(-\omega) = \tilde{T}^*(\omega)$, it follows that if there is a zero at $\tilde{\omega}_k$ then there is also going to be one at $-\tilde{\omega}_k^*$. The zeros appear in pairs which are mirrored on the imaginary frequency axis.
b. Let $z\,(z_r, z_i)$ and $-z^*\,(-z_r, z_i)$ be such a pair. Then their phase contribution to the Blaschke phase is given by [20]:

$$\theta_B = \tan^{-1}\frac{4\omega z_i\left(\omega^2 - |z|^2\right)}{\omega^4 + |z|^4 - 2\omega^2 z_r^2 - 6\omega^2 z_i^2}$$

and this must be added to the minimal phase; similarly, for any other pairs.
c. Distant zeros, i.e. zeros that are far away from the range of frequency measurements, make little contribution to the phase; it is only the near zeros that matter.

A more extensive analysis of the role of zeros can be found in [17, 18, 20, and 21].

It is easy to come up with profiles where the minimal phase is adequate and with others where it is not. Fig. 3.3 shows two such examples. If the profile is something like an asymmetric Gaussian (asymmetric in this context means that the leading and trailing halves of the bunch have different values of σ), then the minimal phase is the true phase and the reconstruction is very good (fig. 3.3a). If, on the other hand, the profile consists of the superposition of two Gaussians, it is relatively straightforward to calculate analytically any zeros in the complex plane and then work out the Blaschke contribution. The existence of any such zeros depends on the full width at half maximum (FWHM), the relative height and the separation of the two Gaussians. Fig. 3.3b is such an example: the assumed profile consists of the superposition of two Gaussian curves, the first with a FWHM = 0.05 ps and the second with FWHM = 0.5 ps (black solid line) that are assumed to be 1 ps apart. The minimal phase on its own would give a rather poor reconstruction (red dot-dash line) but if the Blaschke contribution were to be added (blue dash line) the reconstruction would be excellent. In these two examples, values of ρ were generated for a total of 1000 frequencies. These values were then used to recover the minimal phase according to eq. (3.5), without any additional interpolation or extrapolation; the Blaschke phases were calculated analytically and were added to the minimal phase, when required.

These examples have no great practical significance since in any experiment the existence of Blaschke phases cannot be known in advance, but they do provide some indication about the level of uncertainty that is inherent in any KK-based reconstruction of the bunch profile.

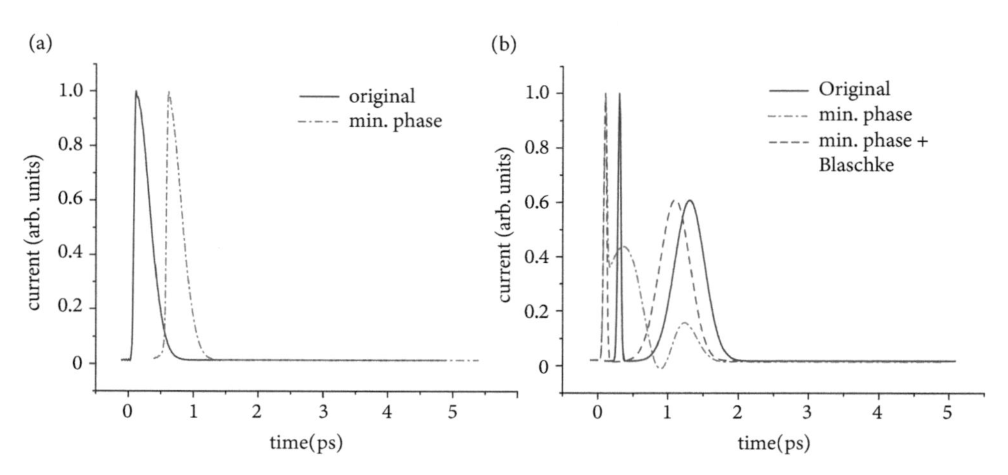

Fig. 3.3 Simulated reconstructions of bunch profiles. (a) An asymmetric Gaussian shape, where the minimal phase (red, dot-dash line) gives an excellent reconstruction of the original profile (black, solid line), and (b) a case consisting of the superposition of two Gaussian curves where the minimal phase, on its own, is inadequate; the addition of the Blaschke contribution (blue, dashed line) recovers the original profile. For reasons of clarity the original and reconstructed curves are shown displaced along the time axis.

3.5 Interpolation and extrapolation

Equation (3.5), for the determination of the minimal phase, implies that ρ is known over all frequencies. Since the set of available values for ρ is invariably restricted to a limited range of frequencies, it is always necessary to interpolate between the available data and to extrapolate beyond. The extrapolation to low frequencies (long wavelengths) is particularly important because, as mentioned earlier, these are the frequencies that determine the overall length of the reconstructed profile. This is done by means of a simple exponential which matches the value ρ_l at the lowest measured frequency ν_{min}. A suitable function for this purpose would be:

$$\rho = \rho_l exp\left[-a\nu^2\right]$$

where ν is the frequency and the parameter a is defined by:

$$a = \frac{-ln\rho_l}{\nu_{min}^2}$$

It should be noted that for a charge normalized distribution, $\rho \to 1$ when $\nu \to 0$. Both this condition and the one requiring $\rho = \rho_l$ when $\nu = \nu_{\mathrm{min}}$ are satisfied by the above function. As far as the high frequency extrapolation is concerned, the requirement is that $\rho \to 0$ for $\nu \to \infty$, as discussed in the previous chapter. This can be satisfied by a function of the type:

$$\rho = \rho_h\left(\frac{\nu_{max}}{\nu}\right)^4$$

where ρ_h is the value of ρ at the highest measured frequency ν_{max}, or by an exponentially decaying function or by a Gaussian; the choice of function for the high frequency extrapolation is not critical.

3.6 Algorithmic phase retrieval methods

Instead of using the KK procedure to recover the minimal phase, it is also possible to derive the missing phase by iterative algorithms which are, essentially, very sophisticated trial and error methods. The basic feature of these algorithms is a series of calculations that alternate between the time and frequency domains. One starts with an initial estimate of the unknown time profile; a forward FT yields the corresponding ρ and θ values in the frequency domain. The values of the phase angle (θ) are retained but the ρ values are discarded and replaced with those derived from the experiment. An inverse FT then gives a second estimate of the profile in the time domain. This is made to satisfy the necessary constraints (or 'supports') in the time domain and the procedure is repeated again and again until the process, hopefully, converges to a solution. This may sound straightforward but there are difficulties, the most common being the tendency to 'stagnation' and very slow convergence. There are a number of variants of this basic process that differ in the constraints that are imposed in each domain. In the frequency domain, the obvious constraint is that the ρ values must match those of the experiment. Additionally, one may combine the algorithm with KK and impose a second constraint, namely that the phase is not smaller than the minimal phase [26]. In the time domain, the recovered function must be real, positive, and non-zero within a finite time window. For a charge-normalized distribution, the integral of the time profile must also tend to unity. These algorithms have been tested, with good results, against hypothetical profiles consisting of combinations of Gaussian or Lorentzian shapes and have been used to analyse experimental data [22–26]. However, the question marks associated with all such inverse problems are still present: how do you ensure convergence? What are the criteria for deciding on the best fit? Is this the only solution? What is the effect of the experimental uncertainties of the measured ρ values on the solution? Two examples of the application of these iterative algorithms to experimental data are shown in fig. 3.4.

The one in fig. 3.4a is taken from [26] and shows the temporal profile of an electron bunch measured at 20.35 GeV. The minimal phase reconstruction (blue, dashed line) is compared to the one derived from an iterative algorithm that constrains the phase to be no smaller than the minimal phase (red, solid line). The agreement is reasonable, although the iterative solution gives a smoother profile. The un-physical negative currents that appear in the tail of the profile recovered by the minimal phase process can only be ascribed to the errors in the process. A second, similar comparison is shown in fig. 3.4b, which is taken from [24]. The beam energy in this case was much lower (15 MeV), and the iterative algorithm was different, relying on a

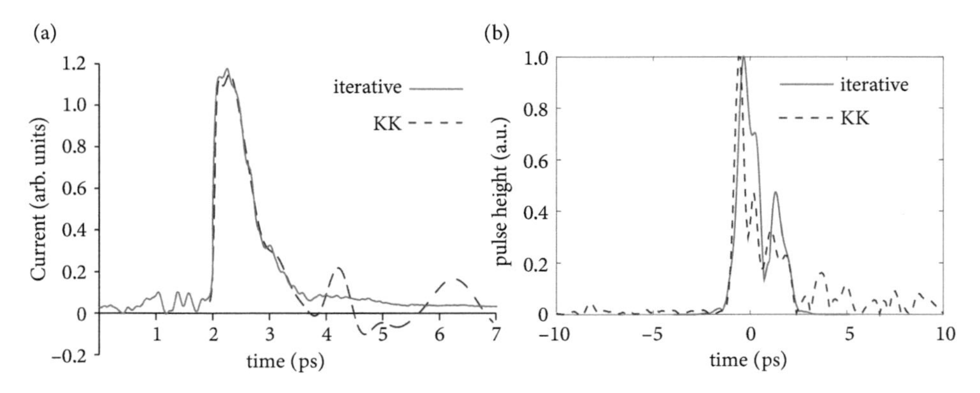

Fig. 3.4 Comparison of the profiles reconstructed from experimental data by the minimal phase (KK) process and by means of an iterative algorithm. (a) taken from ref. [26], and (b) from ref. [24]. In both plots the KK solution is shown with the dashed line and the iterative one with a solid line. See text for further details.

two-stage process whereby a number of phase retrieval runs are then sorted out by means of a cross-correlation analysis. The previous comments are also applicable to this case: the agreement is reasonable, but the iterative method gives a smoother solution.

References

1. K. Yokoya and P. Chen, 'Beam-beam Phenomena in Linear Colliders', KEK Preprint 91-2, April (1991)
2. M.C. Downer et al, Rev. Mod. Physics **90** (2018) 035002
3. M. Uesaka et al, Nuclear Instruments and Methods in Physics Research A **406** (1998) 371
4. C.P. Welsch et al, 2006 JINST **1** P09002
5. K. Nambu et al, Particles 1 (2018) 305
6. V. Dolgashev et al, Phys. Rev. Accel. & Beams **17** (2014) 102801
7. D. Xiang and Y. Ding, SLAC-PUB-14100 (2010)
8. D.X. Wang, G.A. Krafft, and C.K. Sinclair, Phys. Rev. E **57** (1998) 2283
9. K.N. Ricci and T.I. Smith, Phys. Rev. Accel. & Beams **3** (2000) 032801
10. K.N. Ricci, E.R. Crosson, and T.I. Smith, NIM-A **445** (2000) 333
11. I. Wilke et al, Phys. Rev. Letters **88** (2002) 124801
12. G. Berden et al, Phys. Rev. Letters **93** (2004) 114802
13. A.L. Cavalieri et al, Phys. Rev. Letters **94** (2005) 114801
14. G. Berden et al, Phys. Rev. Letters **99**, (2007) 164802
15. A.D. Debus et al, Phys. Rev. Letters **104** (2010) 084802
16. J.D. Jackson, 'Classical Electrodynamics', John Wiley & Sons (1975)
17. J.S. Toll, Phys. Rev. **104** (1956) 1760
18. R. Lai and A.J. Sievers, Nuclear Instruments and Methods in Physics Research A **397** (1997) 221
19. F. Wooten, 'Optical Properties of Solids', Academic Press (1972)
20. O. Grimm and P. Schmüser, TESLA FEL Report 2006-03
21. H.M. Nussenzveig, Journal of Math. Physics **8** (1967) 561
22. S. Marchesini et al, Phys. Rev. B **68** (2003) 1401010 (R)
23. S. Bajlekov et al, Phys. Rev. Accel. & Beams **16** (2013) 040701
24. D. Pellicia and T. Sen, NIM-A **764** (2014) 206
25. M. Heigoldt et al, Phys. Rev. Accel. & Beams **18** (2015) 121302
26. F. Bakkali-Taheri et al, Phys. Rev. Accel. & Beams **19** (2016) 032801

4

Beam Diagnostics with Smith–Purcell Radiation

It will not have escaped the reader's notice that nothing in the previous chapter is specific to SP radiation. Any radiative process can be used, or has been used, for the determination of the time profile of short bunches of electrons (or other charged particles). This is particularly true for transition and synchrotron radiation [1–5], but also for SP radiation [6, 7]. The purpose of this chapter is to highlight the specific advantages of SP radiation. Rather than repeating the details of recent experiments with SP radiation, which can be found in the cited references, it is more profitable to attempt a broader overview of these measurements.

4.1 Sampling points

If one were to summarize in a sentence the main advantage of SP over other radiative processes, one might say that the SP process provides its own, in-built spectrometer. It is also non-destructive because it is not necessary to intercept the particle beam. It may be recalled from the previous chapter that the reconstruction procedure starts from a set of measurements in the frequency domain that yield a set of values for the magnitude (ρ) of the Fourier transform (FT) of the bunch time profile. The simplest approach to the creation of this set of measurements would be a single detector that could slide along an arc around the grating and measure the radiated energy at that angular position, i.e. wavelength. This is certainly possible but in certain cases it may be advantageous to deploy a number of detectors along this arc so that the spectral yield can be determined simultaneously at a number of different wavelengths. The question then arises as to the optimal number for these sampling points. In the experiments of the Oxford group, 11 equally spaced detectors, positioned over the angular range 40–140^0, were used. Therefore, 11 values of ρ are immediately obtainable. Insertion of another grating, with different period, would give another 11 values of ρ, and using three different gratings, as in the Oxford set-up, a set of 33 values of ρ can be created. There is nothing particularly significant about this number and one could try to fit in more or fewer detectors. In principle, the more frequency sampling points the better, but this can only be achieved by increasing the complexity and cost of the detection system and there may be a point where trying to add more sampling points adds very little to the accuracy of the reconstruction. A related question is what is the best deployment of these detectors: is it better to

Smith-Purcell Radiation. George Doucas, Oxford University Press. © George Doucas (2025).
DOI: 10.1093/9780198951360.003.0005

have them at equal *angular* spacing, as in the above experiments, or might it better to position them so that we have equal *frequency* spacing in the measurements? Or, in fact, any other detector spacing? It should be noted that the equal angular spacing arrangement tends to concentrate the sampling points in the low frequency (long wavelength) part of the spectrum. This may not be a bad choice because it is these wavelengths that provide the necessary information about the bunch length but, conversely, coverage of the high frequencies is sparse and information about any fine structure of the bunch may be lost. The equal frequency spacing arrangement, on the other hand, will tend to cram the detectors very closely together and may be difficult to implement. These issues were discussed in some detail in [8]. The study considered the design of a longitudinal profile detector based on SP radiation and involving the use of three different gratings. It was assumed that the sampling would be done at frequencies determined by a suitably chosen geometric progression; the number of the detectors was the free parameter. A 'standard' bunch, whose longitudinal profile was a simple Gaussian curve, was assumed to be propagating through the detector. The robustness of the reconstruction was then tested against deviations from this 'standard' profile, either in length or in shape. The conclusion is that the optimal number of detectors depends on the expected uncertainty (noise) in the measurements. In a zero-noise environment which, evidently, cannot be achieved in any experiment, good reconstruction would have been achieved with a total of just 5 sampling points. If the noise level is 1%, 11–17 points would be required to achieve similar accuracy and for a noise level of 25%, even 35 detectors would not achieve the zero-noise accuracy. This is just a limited sub-set of all the possible scenarios but it does highlight the importance of reducing the experimental uncertainties.

The next question is how to implement any decision about the optimal number of sampling points. Following the previous cited example of, say, 33 frequency sampling points, one could deploy 11 detectors at predetermined angular positions and insert three different gratings into the beamline. This would give the spectral yield at a total of 33 wavelengths but it implies that one grating would have to be retracted and another one inserted in its place, and so on. This is acceptable in cases where the bunches arrive in the form of a bunch train, with intra-bunch spacing that is much shorter than the response time of the detection system. What is then determined is the average shape of the bunch; there are many cases where this is perfectly acceptable and is exactly what has been done in [7] and other similar experiments. There may be other situations, however, where the bunch spacing is rather long and one would like to know the profile of each individual bunch, one notable example being that of beams produced by laser-driven acceleration. In this case, all the information required for the reconstruction of the time profile would have to be provided by one, and only one, bunch; this is the definition of a 'single-shot' measurement. The feasibility of single-shot measurements has already been demonstrated through the use of transition radiation for the determination of the temporal profile of an electron bunch with a length of just under 10 fs (rms). However, a specially designed and rather complex spectrometer, with three independent arms that covered the wavelength range 250 nm to 11.35 μm had to be used for this measurement [9]. In the case of SP radiation there is no need for an external spectrometer, but single-shot determination of the time profile implies that the three gratings would have to deployed simultaneously [10], and not sequentially (as was done in [7]). This still leaves open the question of how to account for any background radiation.

4.2 Background subtraction

Filtering techniques, which will be discussed in the Appendix, will suppress radiation that lies outside the SP band but will not deal with the more difficult problem of how to account for radiation that lies within the SP band but does not originate from the grating itself. This is a situation that could be of crucial importance for certain experiments or applications, the most obvious example being the reconstruction of the temporal profile of ps and sub-ps bunches of charged particles. Two possible approaches are: (a) the use of a 'blank' grating, and (b) the use of the polarization properties of SP radiation.

'Blank' gratings have been used by the Oxford group in a number of experiments [7, 11]. The idea is to replace the grating with an identical piece of metal, but without any corrugations on its surface, hence the name 'blank'. This is inserted near the beam path, in exactly the same position as the grating and a signal is recorded; the difference between the two measurements, one with the grating and the other with the blank, at a given observation angle, might then be thought of as a signal that is due entirely to the existence of corrugations on the metallic surface, i.e. the correct SP signal. If this approach were to be taken, one might then query the need for any filter in front of the detector. This, however, would be wrong for the following reason. Suppose that one were to insert the grating and measure a signal S1, *without* the use of a filter. The signal would consist of the true SP intensity plus a background signal (bck1), i.e.

$$S1 = SP + bck1$$

Then, put in the blank and measure another signal S2, assumed to be the background signal (bck2):

$$S2 = bck2$$

Note that the background must be assumed to be 'white', i.e. covering all wavelengths, and that it might be significantly stronger than the SP component. In an ideal situation bck1 and bck2 would be equal to each other and the true (SP) signal would be:

$$SP = S1 - S2$$

However, bck1 and bck2 are not necessarily exactly the same because the background signal (bck2) is generated without a grating and hence, without a dispersive element which would distribute the wavelengths over the observation angles. Thus, the blank measurement gives a background which may be different from the background generated with the grating because the grating disperses the background. This background fluctuation, which can occur at wavelengths far removed from that of the SP component, can mask the SP signal. Therefore, the use of a suitable filter at each observation angle restricts the measurements to a narrow band around the expected wavelength of SP radiation and reduces the probability of that happening. This reasoning is consistent with the experimental observations.

The use of a blank is appropriate in experiments where one is interested in the determination of the average time profile inside a bunch train, but is obviously not applicable to situations where single-shot measurements are required and the whole process of inserting first a blank, then retracting it and inserting a grating etc. must be avoided. In this case, the exploitation of the polarization properties of SP radiation offers another possibility of discriminating against background radiation, without the use of a blank. The reasoning is as follows. Let the indices 1 and 2 below denote the two orthogonal polarization directions for the SP and the background radiations. The degree of polarization (DOP) α for the SP signal is defined by:

$$a = \frac{s_1 - s_2}{s_1 + s_2}$$

For the background radiation, the DOP β is given by:

$$\beta = \frac{b_1 - b_2}{b_1 + b_2}$$

where s and b are the total SP and background signals, respectively. Therefore,

$$s_1 = \frac{1+\alpha}{2}s \quad s_2 = \frac{1-\alpha}{2}s$$

$$b_1 = \frac{1+\beta}{2}b \quad b_2 = \frac{1-\beta}{2}b$$

Assume also that two measurements have been carried out, one in each polarization direction, and that the two recorded signals are m_1 and m_2, respectively. Then

$$m_1 = s_1 + b_1 = \frac{1+\alpha}{2}s + \frac{1+\beta}{2}b$$

$$m_2 = s_2 + b_2 = \frac{1-\alpha}{2}s + \frac{1-\beta}{2}b$$

Solving this system for s, we obtain

$$s = \frac{m_1(1-\beta) - m_2(1+\beta)}{\alpha - \beta}$$

Hence, it is possible to determine from these two measurements the magnitude of the SP signal, without the need for a blank grating. It is assumed that the two degrees of polarization are different, constant, and known. The penalty in this approach is the need for two detectors per channel, one for each polarization direction, and the consequent increase in the cost and complexity of the detector; however, careful positioning of the three in-line gratings can mitigate some of these problems. One possible deployment of three gratings in order to construct a single-shot bunch profile monitor is shown in schematic form in fig. 4.1 The three gratings are exposed to the particle beam simultaneously but are rotated by 120^0 around the beam axis in order to minimize the overall length of the system (fig. 4.1a). It is assumed that each grating will have 11 observation ports, spaced equally between 40^0 and 140^0 relative to the beam axis. Since background subtraction will be done by polarization measurements, each observation port will require a beam splitter and two detectors, one for each polarization direction; this is shown schematically in fig. 4.1b for just one of the three gratings. This figure is taken from ref. [10] where the reader can find further details of this design.

The DOP of the background signal is site-specific and has to be determined by an independent measurement; it must also be assumed (and verified by repeated measurements) that it is a relatively constant quantity. The DOP of the SP signal has also to be determined experimentally or calculated from the theory, if one is sufficiently confident about this.

4.3 On the origins of background radiation

As mentioned earlier, it is very likely that much of the background radiation could be due to sources of transition or diffraction radiation that may be some distance away from the gratings. This is a site-specific problem that has to be assessed separately for each experiment. There is, however, one source of background radiation that does allow at least an order of magnitude calculation, and this is the heating of the grating itself. Every beam is surrounded by a halo and since there is a strong incentive to bring the beam as close to the grating surface as possible in order to ensure strong coupling, it is inevitable that a small fraction of the beam will be intercepted by the grating, causing the latter to heat up. The grating then becomes a thermal radiator. The wavelength distribution of such a radiator is determined by Planck's law:

$$\Delta P_\lambda = \frac{2hc^2}{\lambda^5} \frac{1}{e^{\frac{hc}{\lambda kT}} - 1} \tag{4.1}$$

(a)

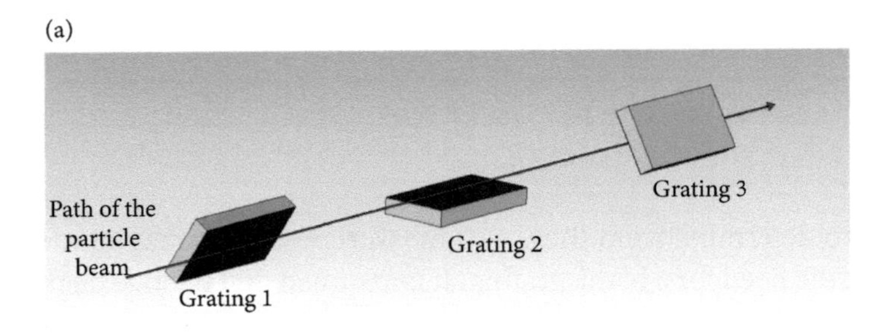

(b)

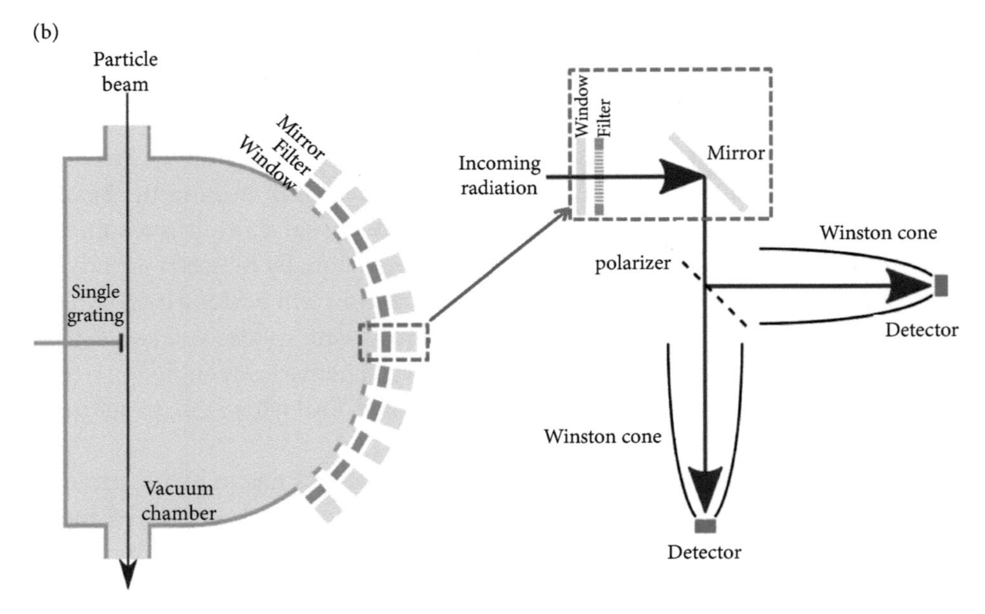

Fig. 4.1 (a) Schematic of three gratings deployed simultaneously along the beam path and (b) schematic of the light collection system for one grating. Adapted from ref. [10].

which gives the emissive power per unit area and per unit solid angle at a wavelength λ and in the interval $d\lambda$; the other symbols have their usual meaning. From Planck's law it is possible to derive the expression for the position of the maxima:

$$\lambda_{max}T = 2898 \ (\mu m.deg)$$

and the total radiated power per unit area (Stefan–Boltzmann law):

$$P = \sigma T^4$$

where the constant $\sigma = 5.67 \times 10^{-8}$ $W.m^{-2}.K^{-4}$

Fig. 4.2 is the graphical representation of eq. (4.1) for black bodies at three plausible temperatures in the context of the present discussion. The grating is bound to be in close thermal contact with its support structure and there would be ample warnings about any significant interception of the beam by the grating, especially

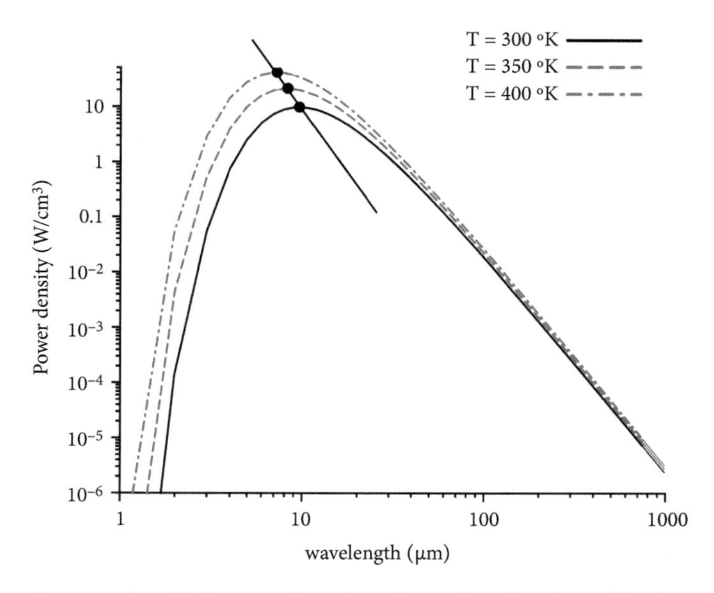

Fig. 4.2 Planck's law as a function of wavelength, plotted for three temperatures. The three dots connected by the straight line are the maxima of the corresponding distributions.

for high energy beams. Therefore, the temperature of the grating is unlikely to be much above 100 ^{0}C.

The peak of the radiation spectrum for a black body at 100 ^{0}C is around 10 μm and in fact, there is extensive overlap between the blackbody spectrum and that of the coherent SP radiation that will be relevant for longitudinal profile diagnostics. Should one be worried about this? Probably not, because according to the above assumptions, a grating with surface area 4×2 cm^2 and at a temperature of 100 ^{0}C would radiate about 1 W, assuming an emissivity of unity. In practice, the infrared emissivity of a metallic grating is going to be below 0.5. Moreover, we are primarily concerned with ultra-short and high charge density bunches and therefore the power emitted in the form of coherent SP radiation is likely to be much higher than 1 W.

4.4 Reconstructed bunch profiles

There is a subtle difference between the determination of the time *profile* of a charged particle bunch and the determination of the *length* of the bunch. The former is the more demanding task because it requires a more detailed analysis of the spectral yield, as described in this and the previous chapters. The determination of the length is somewhat simpler and can also be carried out by measurements in the frequency domain. The principle is straightforward: the onset of coherence, and the consequent dramatic increase in radiated energy, occurs when the wavelength of the radiation is about the same as the bunch length. This can be determined again by the spectroscopic analysis of a radiative process. Coherent SP radiation has been used with excellent results, for both these applications. The reconstruction of the longitudinal (time) profile of picosecond and sub-ps long electron bunches has been carried out at energies as low as 100 keV [12] and as high as 28.5 GeV [11]. An example of the time profile of an electron bunch, derived by analysis of coherent SP radiation at a beam energy of 20.35 GeV, is shown in fig. 4.3.

The oscillations at the tail end of the bunch to negative values are obviously unphysical and are due to the inaccuracies of the experiment or of the Kramers–Kronig (KK) reconstruction, or both. Similarly, the very steep rise at the head of the bunch is an artefact of the minimal phase procedure. This particular profile was reconstructed from measurements taken with three different gratings, with periodicities of 0.05, 0.25, and 1.5 mm. The measured spectral yields are shown in fig. 4.4, together with the rather large experimental uncertainties associated with this particular experiment. Also shown, with solid lines, in this figure are the expected yields from these gratings, calculated according to the surface current theory outlined in the previous chapters.

The reason for this is in order to make a more general and important point. The recovered profile of fig. 4.3 can be fitted by the superposition of two Gaussian curves. This can then be fed into the analysis codes in order to derive the expected yields from these three gratings. It is worth noting that the measured yields *in the long wavelength region* are in reasonable agreement with the expected ones, always bearing in mind the difficulties of precise measurements in the far infrared. When the wavelength is longer than the bunch length, then the details of the bunch structure are no longer important because the bunch appears as a lump of charge;

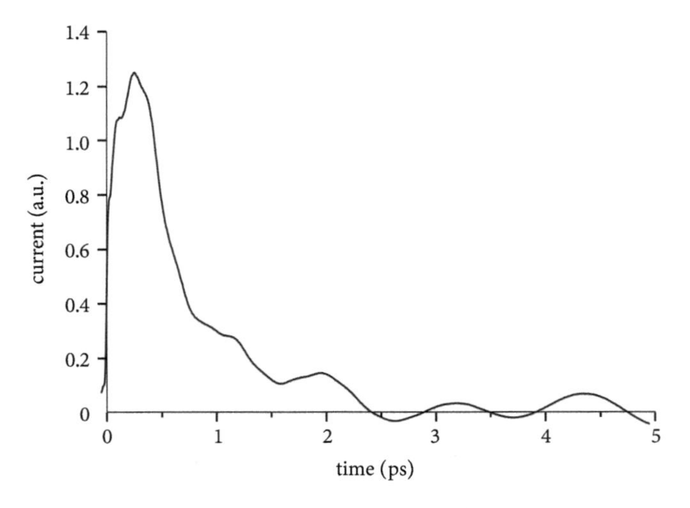

Fig. 4.3 A longitudinal electron bunch profile reconstructed by the analysis of coherent Smith-Purcell radiation; the beam energy was 20.35 GeV (adapted from ref. [7]).

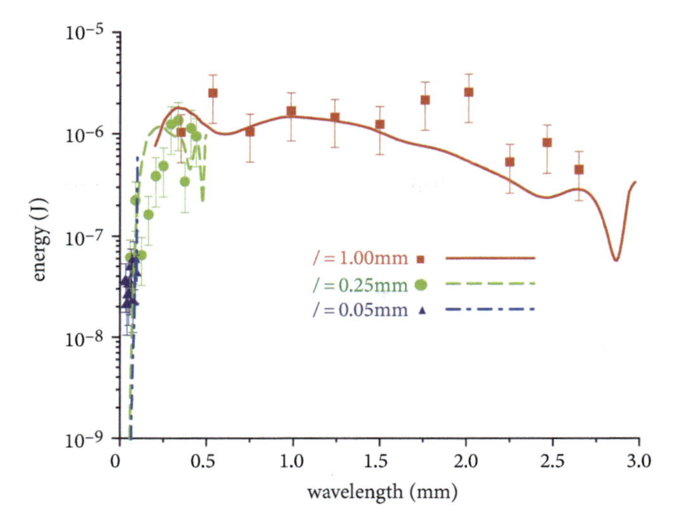

Fig. 4.4 The measured spectral yields (with error bars) for the profile of fig. 4.3. The solid lines are the expected yields for each of the three gratings (1.0, 0.25, and 0.05 mm) used in that experiment (adapted from ref. [7]).

mathematically, this means that the form factor of the bunch is of order unity and the measured yields are a good indication of the validity of the theoretical description of the emission process. Hence, it is correct to claim that the 'surface current' theory agrees with the measurements. Evidently, higher precision measurements would be very desirable.

The low energy (100 keV) experiments of reference [12] have shown that it is also possible to determine the overall bunch length by observing the onset of coherence. The inherently non-invasive nature of SP radiation and the relatively simple and inexpensive experimental set-up are two particular advantages. However, there is a word of caution that is applicable to *all* profile reconstructions that rely on measurements in the frequency domain: since the reconstruction is based on the comparison of measured energies at various wavelengths, it is important to account, as accurately as possible, for all the losses incurred at each wavelength; specifically for detectors, if detector A at frequency f_{a} gives a stronger signal than detector B at frequency f_{b}, is that due to higher SP power at f_{a} or is it due to the higher sensitivity of detector A? Therefore, it is essential that all detectors are calibrated at the frequencies that they are set up to detect. The is an essential but non-trivial task.

4.5 Microbunches

There are a number of applications where it is important to compress the electron beam into bunches of short duration (sub-ps) and of high brightness. Notable examples are the excitation of the very high accelerating gradients in a plasma that are necessary for the development of plasma wakefield accelerators (PWFA) and the generation of ultra-short X-ray pulses in Free Electron Lasers. In any case, charged particle bunches that have durations in the fs range will produce coherent THz radiation and are, therefore, of great interest in the context of the present work. An obvious difficulty in achieving the required high degree of compression of the bunch, especially for low energy ($\gamma < 20$) beams, is the Coulomb repulsion force.

It turns out, however, that in most cases it would be advantageous if the relevant physical process were driven not by a single bunch but by a train of very short microbunches. The term 'microbunch' or 'microbunch train' usually refers to a train of, more or less, identical bunches that are each of very short duration (a few fs) and have a fixed periodicity in the ps or sub-ps range. In actual fact, it is not always necessary for the bunches to be identical or for the periodicity to be fixed. A good summary of the situation can be found in [13] and the references therein. There are a number of techniques that can be used to create the microbunch train, but discussion of this is outside the scope of the present work. What is of interest here is the use of coherent SP radiation as a diagnostic tool for the determination of some properties of a microbunched beam.

The main points of the discussion can be clarified by considering the case of just two microbunches; it is assumed that the microbunches are identical and that their spacing is fixed. Starting from the definition of the FT $F(\omega)$ of a time profile $f(t)$:

$$F(\omega) = \frac{1}{\sqrt{2\pi}} \int_{-\infty}^{\infty} f(t)\, e^{-i\omega t} dt$$

it is easy to confirm that if $f(t)$ is a Gaussian function with amplitude a_0 and standard deviation σ, i.e. if

$$f(t) = a_0 e^{-\frac{t^2}{2\sigma^2}}$$

then its FT is also a Gaussian:

$$F(\omega) = a_0 \sigma e^{\frac{-\omega^2 \sigma^2}{2}}$$

We are interested in charge-normalized distributions, i.e. distributions where the integral of $f(t)$ over all time is equal to unity. In this case, there is a connection between amplitude and standard deviation:

$$a_0 \sigma = \frac{1}{\sqrt{2\pi}}$$

Consider next a time profile consisting of two Gaussian microbunches (a_1,σ_1) and (a_2,σ_2), separated by a time interval Δt. The FT of this profile is:

$$F(\omega) = a_1 \sigma_1 e^{\frac{-\omega^2 \sigma_1^2}{2}} + a \sigma_2 e^{\frac{-\omega^2 \sigma_2^2}{2}} e^{-i\omega \Delta t}$$

Charge normalization now requires that:

$$(a_1\sigma_1 + a_2\sigma_2) = \frac{1}{\sqrt{2\pi}}$$

If we also assume that the charge in the original single bunch has been split evenly between the two identical microbunches, then:

$$F(\omega) = \frac{a_0}{2}\sigma e^{-\frac{\omega^2\sigma^2}{2}}\left(1 + e^{-i\omega\Delta t}\right)$$

Therefore, the magnitude ρ of this FT is:

$$\rho \equiv |F(\omega)| = \frac{1}{\sqrt{2\pi}}e^{-\frac{\omega^2\sigma^2}{2}}cos\frac{\omega\Delta t}{2}$$

which is the same as the magnitude of a single Gaussian, apart from the modulation term $cos\frac{\omega\Delta t}{2}$. Fig. 4.5a shows two charge-normalized time profiles, a single bunch ($\sigma = 0.2$ ps, $a_0 = 1.995$, solid line) and two identical microbunches (dashed line), separated by $\Delta t = 1.0$ ps; fig. 4.5b shows the corresponding FTs.

The magnitude (ρ) of the FT of the microbunch train has the same envelope as that of the single bunch but is now modulated with a frequency that depends on the spacing of the microbunches. This can be exploited in order to monitor continuously the microbunch spacing [14], and SP radiation is particularly useful in this respect because of its inherent flexibility in selecting a wavelength. In the example cited here, there will be minima at frequencies of 0.5 THz, 1.5 THz, etc. The first minimum corresponds to a wavelength of 0.6 mm, and assuming that one is dealing with a highly relativistic beam ($\beta \cong 1$), it would be easy to use a grating with period equal to 0.6 mm and position a detector at 90^0; assuming also that all other experimental

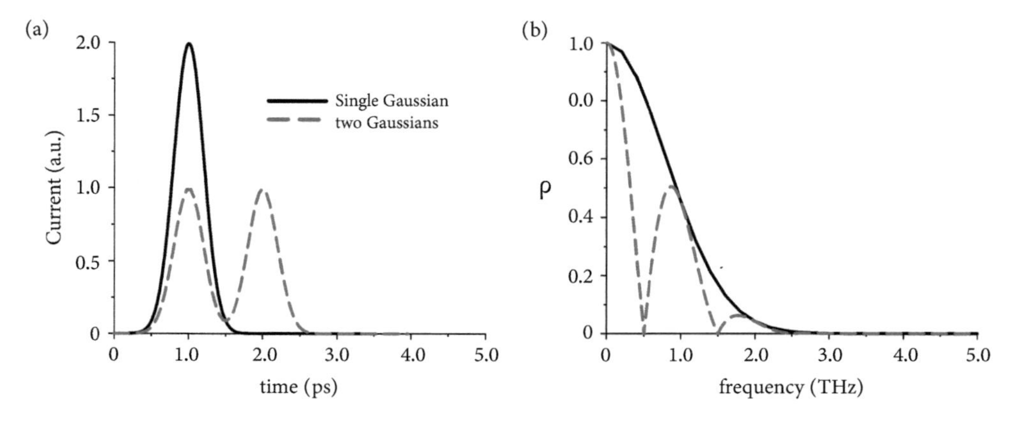

Fig. 4.5 (a) the time profile of a single Gaussian bunch and two identical microbunches, separated by 1 ps; (b) the corresponding magnitudes (ρ) of their Fourier transforms.

conditions are stable, any deviation of the output from its minimal value would be an indication of a change in the bunch spacing.

In the more general case where the charge of a single bunch has been split into M identical microbunches, equally spaced in the train, it can be shown that the FT is:

$$F(\omega) = \frac{a_0}{M}\sigma e^{-\frac{\omega^2\sigma^2}{2}}\left(1 + e^{-i\omega\Delta t} + e^{-i2\omega\Delta t} + \dots + e^{-i(M-1)\omega\Delta t}\right)$$

and its magnitude:

$$\rho = \frac{1}{\sqrt{2\pi}}e^{-\frac{\omega^2\sigma^2}{2}}\frac{\sin\left(\frac{M\omega\Delta t}{2}\right)}{\sin\left(\frac{\omega\Delta t}{2}\right)}$$

For an intra-bunch spacing of Δt picoseconds, the minima would then occur at THz frequencies (ν) given by:

$$\nu = \frac{2k+1}{M\Delta t} \qquad k = \text{integer}$$

A more extensive analysis of microbunches and SP radiation can be found in the work of Sergeeva et al [15].

References

1. M. Heigoldt et al, Phys. Rev. Accel. & Beams **18** (2015) 121302
2. E.B. Blum, U. Happek, and A.J. Sievers, Nuclear Instruments and Methods in Physics Research A **307** (1991) 161
3. Y. Shibata et al, Nuclear Instruments and Methods in Physics Research A **301** (1997) 283
4. G. Schneider et al, Nuclear Instruments and Methods in Physics Research A **396** (1997) 283
5. N.M. Lockmann et al, Phys. Rev. Accel. & Beams, **23** (2020) 112801
6. M.C. Lampel, Nuclear Instruments and Methods in Physics Research A **385** (1997) 19
7. H.L. Andrews et al, Phys Rev. Accel. & Beams **17** (2014) 052802
8. H. Harrison, D.Phil. Thesis, University of Oxford, 2018
9. O. Zarini et al, Phys Rev. Accel. & Beams **25** (2022) 012801
10. I.V. Konoplev et al, Phys. Rev. Accel. & Beams **24** (2021) 022801
11. V. Blackmore et al, Nuclear Instruments and Methods in Physics Research B **266** (2008) 3803
12. P. Heil et al, Phys. Rev. Accel. & Beams **24** (2021) 042803
13. P. Muggli et al, Phys. Rev. Lett. **101** (2008) 054801
14. H. Zhang et al, Applied Phys. Lett. **111** (2017) 043505
15. D.Y. Sergeeva et al, Optics Express **25**(21) (2017)

5

Sources of THz Radiation

The part of the electromagnetic spectrum between 30 μm and 3.0 mm (10–0.1 THz) is usually referred to as 'the THz region'. There is no universal agreement about the exact limits of this region and Bründermann [1], for example, limits it to the range 30μm–1.0 mm; nor is there total agreement about the name itself, although the 'THz region' seems nowadays to be the accepted term. In the past, however, terms like the 'submillimetre' or the 'far infrared' region were also used, depending on whether one viewed this part of the spectrum as an extension of the microwave region or of the infrared region, respectively. As stated earlier, the terms 'far infrared' (FIR) and 'THz' are used interchangeably in this text.

This is a particularly interesting wavelength band because of the many research opportunities and potential applications that it offers. This can be appreciated immediately by inspection of fig. 5.1, which is a simple schematic that transposes the THz frequency band into the corresponding wavelength, temperature, and energy bands.

Thus, in terms of temperature it corresponds to temperatures between 2 K and 450 K, approximately. At its lower limit this is of obvious interest to astrophysics (the cosmic background is at 2.7 K and the interstellar medium at 10 K), while temperatures at the upper limit are easily accessible in a laboratory environment.

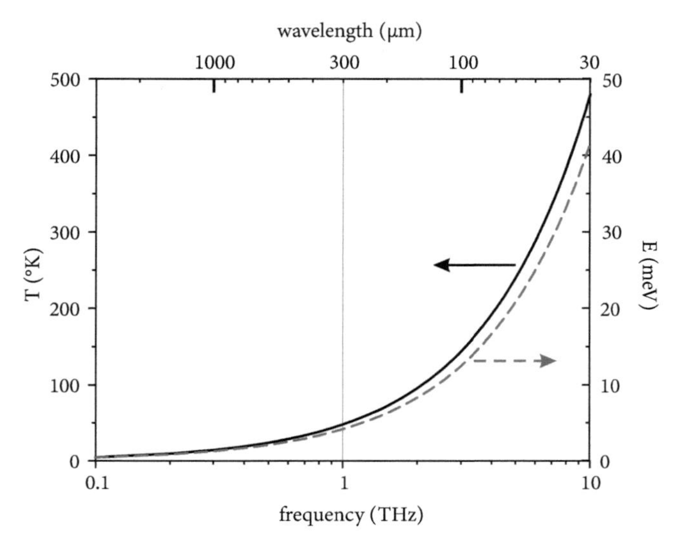

Fig. 5.1 The THz range of the spectrum in terms of temperature, wavelength, and energy.

Smith-Purcell Radiation. George Doucas, Oxford University Press. © George Doucas (2025).
DOI: 10.1093/9780198951360.003.0006

In terms of energy this range covers, for example, rotational and vibrational levels in gases and shallow impurity levels in semiconductors. The former is directly applicable to the remote sensing of various gas molecules in interstellar space and planetary atmospheres. The THz band also coincides with low frequency vibrations of rigid subgroups inside large biological molecules (DNA and RNA) and can be used to determine the structure and topology of these molecules [2]. THz radiation is non-ionizing and therefore understood to be harmless to humans, at least at low power levels. It is also absorbed strongly by both liquid water, and at specific resonant frequencies, by water vapour. These two properties make it attractive to potential biomedical applications and specifically to the non-invasive diagnosis of skin cancers. It turns out that malignant and benign neoplasms have different water content and thus respond differently to THz radiation, which can be used to mark the extent of the malignant neoplasms. This is an area of activity that is still in its early stages and where a number of technical issues still have to be resolved. A recent and comprehensive review of this specific field has been provided by Zaitsev et al [3]. It also happens that many non-metallic and non-polar materials are transparent to, or only weakly absorb at, THz wavelengths. This, combined with the fact that drugs and explosives have characteristic spectra in this spectral region, makes THz radiation interesting for security applications as discussed by Federici et al [4]. One could also add to this list its application to art conservation, e.g. the identification of pigments and binders [5] or the analysis of ancient leather wall papers [6].

The SP radiative process suggests itself immediately as a potential source of tuneable THz radiation and, as mentioned in the Introduction, its most likely application as a source of radiation is going to be in this part of the spectrum. It is, therefore, important to consider briefly the other existing THz sources so that one can appreciate how a grating-based source might fit into the general scheme of things and what advantages it might offer; this is the subject matter of this chapter. There is ample literature on THz sources and techniques and some useful references, where the reader can find more detailed information, are given at the end of the chapter. Moreover, and separate from the potential applications in the THz range, there are very interesting potential applications in the *nm wavelength range*, as will be discussed in the Chapter 7.

All experimental work in the THz region will inevitably require a suitable source of radiation. In certain cases, such as astronomical observations, the source of the radiation is the object under investigation. The term 'THz gap' is often used in the literature in order to indicate that the electronic techniques used in the microwave region have reached their technological limitations and cannot be extended to shorter wavelengths due to the need for very small devices and that, correspondingly, the techniques of the optical region cannot be extended to longer wavelengths due to the comparable magnitude of thermal and transition energies. This does not mean, however, that there are no sources of radiation in this frequency region, as will be clarified below. All sources can be assessed against various criteria, like wavelength range, tuneability, output power in CW or pulsed mode, etc. In addition, it may be necessary in some cases to consider other criteria such as size and operating temperature. Since there is extensive literature on this topic, all that will be presented in this chapter is a broad overview of the subject, with the most basic characteristics of each type of source. The list of sources presented here is not exhaustive but is adequate for the reader to obtain a reasonably accurate picture of the current situation. A detailed discussion of these sources and of the physical principles of their operation can be found in refs [1, 7] and the references therein. A brief but very useful overview of the THz sources can be found in [8].

5.1 Thermal sources

These were the first sources of radiation in the far-infrared part of the spectrum since a black body would be the ideal thermal emitter. The thermal emission of such a body is governed by Planck's law which has been given in Chapter 4 and is repeated below for convenience:

$$\Delta P_\lambda = \frac{2hc^2}{\lambda^5}\frac{1}{e^{\frac{hc}{\lambda kT}} - 1} \tag{5.1}$$

The symbol ΔP_λ denotes the radiance, or the energy emitted per unit surface area and per unit solid angle, in the wavelength interval $\lambda \to \lambda + d\lambda$. The other symbols have their usual meaning. For $hc \ll \lambda kT$, i.e. for high temperature sources and long wavelengths, (5.1) can be approximated by:

$$\Delta P_\lambda = \frac{2ckT}{\lambda^4} \tag{5.1a}$$

whereas for $hc \gg \lambda kT$:

$$\Delta P_\lambda = \frac{2hc^2}{\lambda^5} e^{-\frac{hc}{\lambda kT}} \tag{5.1b}$$

The expression (5.1a) is the Rayleigh–Jeans approximation, while (5.1b) is the Wien one. As the temperature of the black body is increased, the peak of the distribution shifts to shorter wavelengths (Wien's displacement law). Consequently, thermal sources are particularly useful at frequencies above ~3 THz ($\lambda < 0.1$ mm). For the far infrared part of the spectrum, the only thermal sources of practical significance are the Globar, which consists of a rod of SiC heated to about 1500 K, and the mercury arc lamp. In the case of the lamp, the sources of the radiation are the arc discharge itself (5000 K) and also the quartz envelope of the lamp, which can reach temperatures of about 1000 K.

5.2 Gas lasers

Molecular vapour lasers can produce significant levels of THz radiation from inter-rotational transitions in molecules with a permanent dipole moment. In the early years they were excited by means of a glow discharge that is contained in a Fabry–Perot cavity. From the 1970s on they were superseded by the optically excited molecular vapour lasers where the excitation is provided by a pump laser, usually a CO_2 one ($\lambda \approx 10$ μm). These are now a more often used alternative which can provide many discreet wavelengths in the range 0.3–5 THz (1.0–0.06 mm) [7, 9]. However, they are rather bulky devices that consume significant amounts of energy and have very limited tunability. Their output power is between 0.1 and 0.4 W in CW mode and 0.5 and 1.0M W in pulsed mode.

5.3 Semiconductor lasers

Semiconductor lasers are common in the visible or near-infrared part of the spectrum. They are based on a p–n junction of a material like GaAs and the emission of light is due to electron-hole recombination across the band gap. This process, however, does not apply to THz emission because even in the smallest direct band gap materials, which are some lead salts, the gaps correspond to wavelengths that are shorter than ~3–20 μm [20], and therefore are too short to be considered as part of the THz spectrum. In spite of this comment, it is still possible to obtain THz emission from bulk Ge and Si by exploiting different physical processes in these two materials. For our purposes it suffices to note that THz lasing in p-type germanium can be achieved by the simultaneous application of crossed electric and magnetic fields of the order of kV/cm and of a few Tesla, respectively. Further details can be found in the work by Bründermann et al [1]. THz lasing has also been achieved in n-type silicon by optical pumping with a CO_2 or a Free Electron Laser (FEL). These two types of THz source can provide a number of discrete lines over the wavelength range 35–250 μm, approximately [10].

More widespread in recent years is the use of Quantum Cascade Lasers (QCL) [11]. These are periodic structures consisting of many alternating layers of semiconductor materials, typically GaAs/AlGaAs, that form quantum energy wells. Inside the wells the conduction band is split into two or more sub-bands. Under the influence of the external electric field the electron makes a transition from the higher to the lower sub-band, emits a photon and is subsequently extracted to the next period where the process is repeated. Hence, it is possible to obtain a cascade of photons, theoretically one per electron, per period. It is worth noting that QCLs are unipolar devices because, in contrast to other semiconductor lasers where the lasing transitions occur between the conduction and valence bands, in the QCLs the transitions are between the sub-bands inside the conduction band; no holes are involved in the lasing process. A second interesting feature of these devices is the fact that the transition levels, i.e. the emitted wavelength, are determined not so much by the material properties but by the thickness of the layers. Although QCLs are used primarily in the 3–50 μm range, it is possible to extend their operating range into the THz region, approximately between 0.8 and 4.9 THz (0.38 and 0.06 mm). In this case, however, the devices have to be cooled down to cryogenic temperatures. This is an area where considerable research effort is being invested in order to increase the operating temperatures and to extend lasing to longer wavelengths.

It is also possible to generate THz wavelengths by using the well-known process of mixing two different frequencies in a non-linear device, whereby a sum and a difference frequency are generated. By suitable choice of the two frequencies, say from the visible or infrared domain, it is possible to arrange so that the difference frequency is in the THz domain. Sources based on this idea are known as photomixers or frequency difference generators. Alternatively, millimetre waves can be successively frequency-doubled or tripled to reach over 1 THz. This involves passing the output of a microwave signal generator though a non-linear device, a metal–insulator–metal (MIM) diode or, most often, a GaAs Schottky diode. The wavelength coverage provided by this method extends from 0.1 THz into the few THz region, but frequencies up to ~9 THz have also been reported [12]. The CW powers available range from fractions of 1 W at 0.1 THz to less than 1 mW at 1 THz. A useful summary of the performance of the various solid state sources can be found in [13].

5.4 THz radiation from free electrons

One common characteristic of all the previously mentioned sources is that they rely on processes that occur in a body, be it solid or gaseous. There is a second group of sources whose operating principle is based on the behaviour of free electrons in vacuum. These can be further subdivided either into low electron energy devices (a few keV) or into relativistic electron energy devices that operate from a few MeV up to GeV energies. The most relevant example of the first category is the Backward Wave Oscillator (BWO). This is a table-top device where the electron beam, typically 1–10 keV in energy, passes over a periodic structure as it is accelerated towards the anode. The interaction of the beam with the periodic structure results in beam bunching and in the transfer of energy from the electrons to the counter-propagating electromagnetic wave, hence the 'backward wave' name. It is also necessary to apply a strong axial magnetic field in order to collimate the beam. Frequency tuning can be achieved by changing the accelerating voltage. It is possible to deploy a number of BWOs and thus cover the range from 30 GHz to 1.4 THz, approximately [7,8]. The achievement of higher frequencies is, however, problematic because the periodicity of the structure must be of the same order as the wavelength of the radiation. This requires periodic structures with very fine spacing and a general reduction of all the physical dimensions of the device, with the inevitable power dissipation consequences. There are similarities with the Smith-Purcell process in the sense that in both cases the radiation is due to the interaction of an electron beam with a periodic structure, but there are differences as well: the BWOs are rather low energy devices that require an external magnetic field and where the electron beam is confined inside the interaction cavity; none of these apply to SP radiation.

In the second group of free electron sources, those that are based on the acceleration of multi-MeV electrons in vacuum, one has to consider the FELs and the sources based on synchrotron radiation. It is well known that the acceleration of any charged particle (electrons for the purposes of this discussion) will result in the emission of electromagnetic radiation. In the case of the FEL, the electron beam is made to 'wiggle' as it travels between the pole gaps of a periodic array of magnets of alternating polarity. This structure is called an undulator or a wiggler, depending on the strength of the magnetic field (stronger in a wiggler), and is based on the pioneering work of Motz [14] in the early 1950s. For the purposes of our discussion the terms are used interchangeably, but they produce radiation with different characteristics [17]. Mirrors placed at both ends of the wiggler create a low-loss cavity. As the electrons travel through the wiggler they are subjected to transverse acceleration and emit radiation over a range of wavelengths. There is a resonance condition given by:

$$\lambda = \frac{l_u}{2\gamma^2}\left(1 + \frac{K^2}{2}\right) \tag{5.2}$$

which connects the wavelength (λ) of the radiation, the relativistic factor (γ) of the beam and the period (l_u) of the undulator which is defined as the distance over which the magnetic field completes one period. When this condition is satisfied the electrons interact with the electromagnetic wave and begin to form bunches that are separated by one wavelength. It is this bunching that stimulates the emission of coherent radiation at that wavelength. The quantity K in (5.2) is known as the undulator parameter and is defined (*in the CGS system*) by:

$$K = \frac{eBl_u}{2\pi mc^2} = 9.32 \times 10^{-5} B\,(gauss)\, l_u\,(cm)$$

where B is the magnetic field strength and e and c have their usual meaning. Tuning of the emitted radiation is usually achieved by exploiting the strong dependence of the wavelength on the relativistic factor γ. FELs are widely tuneable devices that can extend, at least in principle, from the X-ray to the THz part of the spectrum. However, these are sources that require big accelerator facilities and there are also limits in the range of γ values that are achievable at a given accelerator. Those operating in the far infrared (FIR) region seem to cover the range 3–2500 μm, approximately [7, 8]. A recent list of FEL facilities worldwide can be found in [15].

Instead of using a wiggler, it is possible to obtain FIR wavelengths by exploiting the synchrotron radiation that is emitted when a charged particle's trajectory is altered by a bending magnet. The subject of synchrotron radiation has been discussed in detail in a number of textbooks on electrodynamics [16] and in the context of the physics of particle beams and accelerators [17]. It is worth reminding the reader that transverse acceleration, as in a bending magnet, produces much more intense radiation than the linear one ('linear' implies that the acceleration and velocity vectors are parallel). Specifically, for a given accelerating force (dp/dt), where p is the momentum of a particle of mass m, the radiated power P in the case of linear acceleration is given by:

$$P = \frac{2}{3}\frac{e^2}{m^2c^3}\left(\frac{dp}{dt}\right)^2$$

whereas in the transverse case by:

$$P = \frac{2}{3}\frac{e^2}{m^2c^3}\gamma^2\left(\frac{d\bar{p}}{dt}\right)^2$$

Hence, the radiated power in the latter case is a factor γ^2 stronger than in the former [16]. Thus, for all practical purposes, the radiation emitted by the linear acceleration of a particle can be ignored. The bending magnet is usually part of an electron storage ring but it is also possible to use a linear accelerator to bring the beam to the required energy and then use a magnet to produce the synchrotron radiation. The advantage of the storage ring is the very high repetition rate achieved from a series of circulating electron bunches, (in the 100 MHz range), and hence high average power. At infrared and shorter wavelengths, it is sometimes difficult to produce the very short bunch lengths that are necessary to bring the radiation into the coherent regime and the emitted incoherent power scales as the number of electrons in the bunch. For the THz region, it is sometimes possible to configure the synchrotron to have short enough bunches so that the emitted radiation becomes coherent. In this case, the emitted power is proportional to the square of the number of electrons in the bunch, and useful radiated powers can be obtained over most of the THz range [18–20]. Linear accelerators (linacs), on the other hand, have a rather modest beam repetition rate, of the order of a few Hz, which results in rather low average power, but can produce bunches in the sub-picosecond range. Increasing the repetition rate of a typical linac to that of a storage ring would be prohibitive in terms of power dissipation. However, in recent years the situation has started to change with the advent of the so-called energy recovery linacs (ERL). The use of superconducting RF accelerating cavities and the recovery of the energy of the spent electron beam before it is dumped has made it possible to increase the bunch repetition rate in this type of linac to almost to the level of a storage ring [21, 22].

In summary, both FELs and synchrotron radiation originating from a bending magnet can provide useful output powers over most of the THz spectrum. The bending magnets can be deployed either as elements of a storage ring or as a single magnet at the end of a linac. In the case of a storage rings the challenge is to keep the bunch length short enough to bring to radiation into the coherent regime. In either case the size and complexity of the accelerator installation is a serious limitation.

References

1. E. Bründermann, H.W. Hubers, and M.F. Kimmitt, 'Terahertz Techniques' Springer (2012)
2. T. Globus et al, J. Phys. D: Appl. Phys. **36** (2003) 1314
3. K.I. Zaytsev et al, J. Opt. **22** (2020) 013001
4. J.F. Federici et al, Semicond. Sci. Technol. **20** (2005) S266–S280
5. K. Fukunaga et al, IEICE Electronics Express **4** (2007) 258
6. A. Doria et al, Appl. Sci. **10** (2020) 7661
7. S.S. Dhillon et al, J. Phys. D: Appl. Phys. **50** (2017) 043001
8. G.P. Gallerano and S. Biedron, Proceedings of the 2004 FEL Conference 216
9. G. Dodel, Infrared Physics & Technology **40** (1999) 127
10. H.-W. Hubers, S.G. Pavlov, and V.N. Shastin, Semicond. Sci. Technol. **20** (2005) S211
11. R. Kohler et al, Nature **417** (2002) 156
12. H. Odashima, L.R. Zink, and K.M. Evenson, Optics Letters **24** (1999) 406
13. Virginia Diodes Inc. *https://www.vadiodes.com/images/AppNotes/ApplicationNote-SummaryofSolid-StateSources.pdf*
14. H. Motz, Journal of Appl. Physics **22** (1951) 527
15. P.J. Neyman et al, Proceedings of the 38th FEL Conference (2017) 204
16. J.D. Jackson, 'Classical Electrodynamics', Ch. 14, John Wiley & Sons (1975)
17. K. Wille, 'The Physics of Particle Accelerators', Oxford University Press (2000) and Joint Universities Accelerator School (JUAS) (2013)
18. BESSY II *https://www.helmholtz-berlin.de/forschung/quellen/bessy/index_en.html*
19. M. Abo-Bakr et al, Phys. Rev. Lett. **88** (2002) 254801
20. F. Wang et al, Phys. Rev. Lett. **96** (2006) 064801
21. G.L. Carr et al, J. Biol. Phys. **29** (2003) 319
22. G.P. Williams, Rep. Prog. Phys. **69** (2006) 301

6

The Smith–Purcell Process as a Source of THz Radiation

The potential usefulness of Smith–Purcell (SP) as a source of THz radiation has to be assessed against three criteria: (a) tuneability of the source, (b) ability to cover the whole of the THz region (0.1–10 THz), and (c) radiated power. In addition, one might also want to consider other parameters such as compactness of the installation, cost, etc.

6.1 Tuneability

On the subject of tuneability, it is evident from the basic SP formula that this can be achieved either by varying the energy of the electron beam, i.e. the parameter (β), or the angle of observation (θ), or both. However, even for a moderately relativistic electron beam with kinetic energy of, say, 500 keV ($\gamma \cong 2$) β is already equal to about 0.87 and the scope of tuning the emitted wavelengths by beam energy variation alone is very limited. For electron beam energies up to about 100 keV ($\gamma < 1.2$) it is possible to achieve effective tuning of the emitted wavelength by simply varying the accelerating potential of the beam (see also the Chapter 7). Beyond that, changing the energy has a small effect on the emitted wavelength and tuning would have to be done by changing the observation angle (θ).

6.2 FIR coverage

In most SP experiments observation angles in the range between 30^0 and 150^0 are easily accessible from the experimental point of view, without the need for complicated light-collection systems. Assuming that one is operating in the relativistic regime with $\beta \cong 1$, it is easy to verify that just two gratings with periodicities of 0.25 mm and 1.60 mm, respectively, would be sufficient to cover the entire far infrared (FIR) region, as defined here, namely from 0.1 to 10THz, or in terms of wavelength from 3.0 to 0.03 mm. These are gratings that are easy to manufacture in a precision mechanical workshop. However, this is not meant to imply that one should use *only* two different gratings or that these two periodicities are necessarily the optimal ones. More than two gratings and/or different periodicities may give higher output over a desired wavelength range and it is up to the designer of the

Smith-Purcell Radiation. George Doucas, Oxford University Press. © George Doucas (2025).
DOI: 10.1093/9780198951360.003.0007

experiment to decide on the best configuration. Fig. 6.1 is one of the many possible configurations, based on the use of just two blazed gratings.

Note that if the forward observation angle is restricted to 30^0 and the desired wavelength is about 30 μm, the shortest wavelength under consideration here, then one grating must have a period of 0.25 mm. One might then ask why not use a grating of even shorter period and observe the same wavelength at less shallow angle θ. The answer to this can be found in eq. 1.7a of Chapter 1, which gives the single-electron yield at a frequency ω and which is repeated below:

$$\frac{dI}{d\Omega} = \frac{e^2\omega^3}{4\pi^2 c^3}\frac{Zl}{n}e^{-\frac{2x_0}{\lambda_e}}R^2 \tag{6.1}$$

For a given frequency and a given grating length, the output increases linearly with the grating period (l), assuming of course that the grating efficiency factor R^2 remains constant. This, however, is not necessarily the case and the grating efficiency must be calculated for each grating but, as a general rule of thumb, it is preferable to obtain the desired wavelength from a longer, rather than a shorter, period grating. Moreover, one should also bear in mind the effect of θ on the evanescent wavelength λ_e; according to the discussion of section 1.1a, in the case of relativistic beams ($\gamma \gg 1$) and non-zero azimuthal angle ϕ, λ_e is approximately equal to:

$$\lambda_e \cong \frac{\lambda}{2\pi \sin\theta \sin\phi}$$

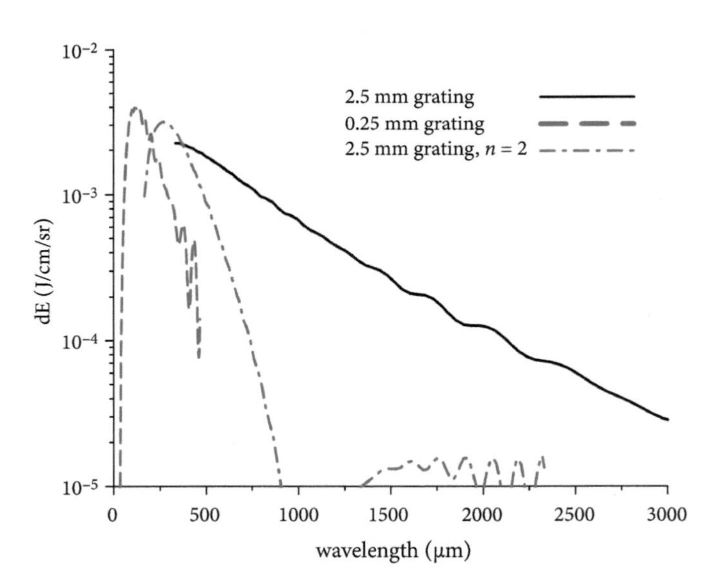

Fig. 6.1 The expected differential energy yield dE (in J per cm of grating length per solid angle) for a set-up consisting of two gratings that can cover the whole of the FIR part of the spectrum. Also shown, by the dot-dash line, is the expected 2nd order yield from the 2.5 mm grating. Further details of the assumptions made in this simulation are given in the text.

Hence, smaller values of θ will increase λ_e and improve coupling between beam and grating. The same comment applies at the other end of the spectrum: a wavelength of, say, 1200 μm could be obtained at $\theta = 60^0$ or 90^0 or 120^0 from gratings with periods of 2400 μm, 1200 μm, and 800 μm, respectively; the longer period grating would be the better choice. The example of fig. 6.1 uses two gratings with periods of 0.25 mm and 2.5 mm, which will cover the entire far-infrared (FIR) region; both of them are assumed to be of the echelette type, with the first blaze angle equal to 20^0. The expected differential energy yield dE (in J/cm/sr) was calculated on the assumption that a short (0.1 ps full width at half maximum (FWHM)) bunch is propagating at a distance of 1.0 mm above the gratings; the bunch charge is 10^{10} electrons and the bunch time profile is assumed to be a symmetric Gaussian. These parameters are plausible for the case of a linear accelerator.

There are a couple of noticeable features in this figure that deserve some comment. The first is that most, but not all, of the FIR spectrum could be covered by just one grating. However, the yield varies widely over this spectral range, almost by two orders of magnitude. This can be mitigated by the use of more than two gratings and is, in fact, a good argument in favour of a system with not two but three or more gratings. The second point is the very rapid drop in output for wavelengths shorter than about 70 μm. This is due entirely to the reduction in the coherence of the emitted radiation. In other words, the emitted energy will depend not only on the choice of gratings but, most importantly, on the intensity and the form factor of the electron beam that is driving the radiative process. In the example chosen for fig. 6.1 the bunch length was assumed to be 0.1 ps (FWHM). If it desired to increase the output at wavelengths shorter than 70 μm, then it would be necessary to compress the bunch further in order to make its length significantly shorter than 70 μm and thus increase the coherence of the emitted radiation.

The output of the simulation of fig. 6.1 is given in terms of energy per cm of grating length and per steradian of solid angle. What is this likely to mean in terms of useful energy? If we assume a grating length of 4 cm and a solid angle of 5 msr, both of which are realistic values, then an output of, say, 10^{-3}J/cm/sr corresponds to an energy of 2×10^{-5}J per electron bunch, in the wavelength region 500–1000 μm. This, however, is a theoretical estimate and one has also to consider the inevitable losses between source and detector. These are likely to be significant and if 10% of the emitted energy reaches the detector, then the useable energy is going to be of the order of one μJ. This is what can be realistically expected from a single 0.1 ps bunch of 10^{10} electrons propagating 1.0 mm above these gratings.

6.3 Intensity of radiation

In a previous chapter an attempt was made to compare the intensities of SP radiation and transition radiation (TR). Since synchrotron radiation (SR) is another radiative process that is based on free electrons and is widely used in the THz part of the spectrum, it would be interesting to carry out, in a similar fashion, a rather basic comparison between the intensities expected from SR and SP radiation. The comparison will be made on the basis of the single-electron yields and at a wavelength of 1.0 mm, i.e. at 300 GHz. The spectral yield of synchrotron radiation at an angular frequency ω, which is far lower than the critical frequency (see below for the definition), and within a bandwidth $d\omega$ is given by:

$$\frac{d^2I}{d\omega d\Omega} \cong \frac{e^2}{c}\left[\frac{\Gamma\left(\frac{2}{3}\right)}{\pi}\right]^2\left[\frac{3}{4}\right]^{\frac{1}{3}}\left[\frac{\omega\rho}{c}\right]^{\frac{2}{3}} \tag{6.1}$$

where ρ is the radius of curvature of the particle's trajectory [1] and the symbol Γ denotes a gamma function. Putting in the numerical factors, the above expression becomes:

$$\frac{d^2I}{d\omega d\Omega} \cong 0.17\frac{e^2}{c}\left[\frac{\omega\rho}{c}\right]^{\frac{2}{3}} \tag{6.1a}$$

The critical frequency (ω_c) divides the spectrum in two parts of equal intensity and is defined as:

$$\omega_c = 3\gamma^3\frac{c}{\rho}$$

Note that the brackets in the above expressions are dimensionless and that the units of the yield are erg.s; note also that in some papers [2] the critical frequency is defined as half the value given above. For a radius of curvature of 100 cm, an energy of at least 25 MeV ($\gamma = 50$) would be required in order to ensure that the critical frequency is above the THz part of the spectrum. Since most synchrotron radiation sources run at much higher energies, the use of 6.1a above is appropriate. For the purposes of this study, it is assumed that the beam energy is 50 MeV ($\gamma = 100$).

In Chapter 1 (equations 1.7 and 1.7a) it was shown that the single electron SP yield at a frequency ω and within a bandwidth $d\omega$ from a grating having N periods is given by:

$$\frac{d^2I}{d\omega d\Omega} = \frac{e^2\omega^2N^2l^2}{4\pi^2c^3}e^{-\frac{2x_0}{\lambda_e}}R^2 \tag{6.2}$$

In order to facilitate comparison with the corresponding expression for the synchrotron radiation yield, this can be rewritten as:

$$\frac{d^2I}{d\omega d\Omega} = \frac{1}{4\pi^2}\frac{e^2}{c}\left[\frac{\omega lN}{c}\right]^2 e^{-\frac{2x_0}{\lambda_e}}R^2 \tag{6.2a}$$

where the quantity in the brackets is again dimensionless. We ignore, for the moment, the exponential term and the grating efficiency factor R^2 and we compare the yields in units of e^2/c. It is assumed that the synchrotron radiation is generated by the electron travelling on a radius of curvature of 100 cm and that the grating has a period $l = 1.0$ mm and a total of $N = 40$ periods. According to (6.1a) the synchrotron radiation output would be about 58 units, while from (6.2a) that of SP radiation would be 1600 units. We now have to provide realistic values for the exponential term and for R^2 in the expression for SP radiation (6.2a). As far as the exponential term is concerned, it is reasonable to assume that the beam can be brought to a distance $x_0 = 1.0$ mm above the grating surface and that the azimuthal acceptance of the light collection system will be of the order of $\phi = 3^0$, in which case the evanescent wavelength (λ_e) will be approximately 3 mm; hence, $e^{-2x_0/\lambda_e} \cong 0.52$. The grating efficiency factor depends on the design of the grating. For a strip grating it would be between 0.1 (eq. 1.9) and 0.02 (see Chapter 1, fig. 1.7), depending on the way this type of grating is treated in the simulations. For a more realistic blazed grating with a 30^0 blaze angle, the numerical calculations yield values between 0.3 and 0.4 in the azimuthal range $0 < \phi < 3^0$. Combining these numbers together, we get an estimated yield of about 290 units for SP versus 58 units for synchrotron radiation.

It is possible to do this comparison in another way, using the parameters given by Williams in ref. [3] which discusses the properties of synchrotron radiation from an experimentalist's perspective. A bunch with duration of 0.1 ps (FWHM) is circulating in the ring with a frequency of 100 MHz. The bunch charge is 100 pC, i.e. there are about 6×10^8 electrons in the bunch. According to the above reference, the synchrotron radiation yield per unit wavenumber at a wavelength of $\lambda = 1000$ μm and in a collection angle of $150 \times 150(\text{mrad})^2$ is, approximately, 5 Wcm. In terms of wavelength, this becomes $\frac{dP}{d\lambda} = \frac{5}{\lambda^2} W/cm$. In order to compare with SP radiation, we need to assign a bandwidth $d\lambda$ that is the same as that expected from the grating; for the previously described grating with $N = 40$, this is 25 μm and using this value for the synchrotron radiation estimate, we obtain 1.25 W. This is the expected power as synchrotron radiation at a wavelength of 1000 μm and in a bandwidth of 25 μm, from a bunch with the previously stated parameters and circulating in the ring at a frequency of 100 MHz. If one now runs a simulation code for the SP radiation, the expected yield is about 3×10^{-8} J *per bunch*, in a solid angle of about 6.5 msr which is smaller than the one assumed for the synchrotron radiation simulations; therefore, the SP radiation radiated power, in this scenario, would be about 3 W. These numbers must be seen as indicative only and one could play with them ad infinitum but the conclusion that can be drawn from this exercise is that the yield from the grating *in the THz part of the spectrum* is going to be at least as good, and probably significantly better, than that from a bending magnet.

6.4 Higher-order emission

The subject of higher-order emission was introduced in Chapter 2 but the discussion there was centred around the single electron yield expected from various emission orders and as such, it is applicable only to cases where the electron beam that excites the emission of radiation is DC, or sufficiently dilute (i.e. very long bunch length) so that coherence effects are negligible. Of potentially greater practical interest are situations where the electron beam is tightly bunched and coherence dominates. An example of this is shown in fig. 6.1 above, where the 2nd order yield from the 2.5 mm period grating is compared with the 1st order emission. In the wavelength region where there is an overlap between the two emissions, 1st order emission is clearly more intense. There is, however, a narrow region between 200 and 380 μm, approximately, where would it would be advantageous to opt for 2nd order emission from the 2.5 mm grating, rather than 1st order from the 0.25 mm one. Higher-order emission, therefore, deserves some consideration.

This is just one example of what could be achieved with the use of two gratings but it is not meant to imply that these are necessarily the optimal gratings. Significantly higher outputs would certainly be obtained by bringing the beam closer to the grating surface and possibly by changing the blaze angle of the grating. The discussion in this section was centred around the question of how many gratings would be required to cover the entire FIR part of the spectrum. It is very likely that there may be applications where the emphasis is not so much on the breadth of the coverage but on the intensity of the radiation over a limited sub-range of the FIR. In all cases, the choice of grating(s) would have to be decided by simulations that are aimed at the requirements of that specific application.

References

1. J.D. Jackson, 'Classical Electrodynamics', Ch. 14, John Wiley & Sons (1975)
2. K. Wille, 'The Physics of Particle Accelerators', Oxford (2000) and Joint Universities Accelerator School (JUAS) (2013)
3. G.P. Williams, Rep. Prog. Phys. **69** (2006) 301

7

Smith–Purcell at Low Beam Energies and Short Wavelengths

Up to this point the reader may have been left with the impression that the Smith–Purcell (SP) process, and its exploitation, require relativistic beams and metallic gratings and that the radiated wavelengths will be in the far-infrared (FIR) part of the spectrum. None of these assumptions are necessary, as will be discussed in this chapter.

The reason for revisiting these rather basic assumptions is that the wavelength range around 1 μm is of great importance in the optical communications industry because of its low attenuation rate in the optical fibres. A tuneable source of near-infrared radiation, based on free electrons, would be of great interest; hence, a significant amount of effort has been invested in recent years in exploring the possibilities offered in this region by SP radiation [1, 2]. The whole set-up for such an application would be very different from what has been discussed so far in this book and it is worth taking each feature of SP radiation in turn in order to consider its possible extension to the near infrared.

The first observation to make is that any application in the communications industry will not require large installations and multi-MeV beams. On the contrary, the electron beam energies are going to be in the few keV range and the device is going to be a table-top one. Rather than changing the observation angle, as in the case of relativistic beams, tuning could be achieved easily by varying the electron beam energy. The emitted radiation would be collected at a fixed angle, probably at 90^0. This works very well for low energy beams ($\gamma < 2$ or $\beta < 0.86$, approximately) but for higher energies this method of tuning becomes inefficient because the relativistic velocity β is approaching unity. For very low energies it is helpful to introduce a parameter ε which is the ratio of the beam energy E_k over the electron mass ($\cong$ 511 keV), i.e. $\varepsilon = \gamma - 1$. Fig. 7.1 shows the emitted wavelength λ (expressed as a fraction of the grating period l) as a function of the parameter ε. This is shown for three observation angles. It is worth noting the steepness of the tuning curve for E_k < 10 keV, approximately.

Wavelengths of, say, 1500 nm would require gratings with periodicities in the nanoscale range. Such gratings can be manufactured routinely on a silicon substrate, maybe not as blazed gratings but certainly as what has been termed 'lamellar' type ones. Bearing in mind the importance of the evanescent wavelength λ_e in achieving close coupling between beam and grating (see Chapter 1), it would be essential to have a low emittance beam that can be brought close enough to the grating, a

Smith-Purcell Radiation. George Doucas, Oxford University Press. © George Doucas (2025).
DOI: 10.1093/9780198951360.003.0008

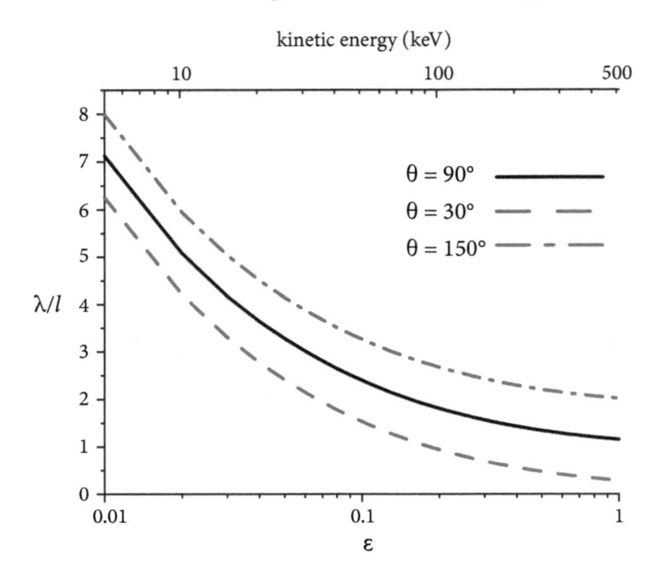

Fig. 7.1 Tuning the wavelength by means of an energy change of the electron beam. The ratio (λ/l) of the emitted wavelength to the grating period is plotted against the parameter ε for three observation angles. The electron beam energy is shown on the top axis.

separation again in the nanometre range. This can be done by using the beam from a scanning electron microscope (SEM) [3], but it may also be possible to incorporate a silicon field emitter source on the silicon chip itself and thus achieve a tuneable, all-silicon source of micron or even sub-micron radiation.

The more fundamental and interesting question in the context of the present work is whether it is possible to produce SP radiation from a semiconductor grating, say silicon, and to provide a simple physical picture for the origin of this radiation, as well as an analytical (or semi-analytical) solution for the calculation of the radiated energy. Extensive research in recent years confirmed that SP radiation can indeed be produced from Si gratings using electron beams with energies as low as 2 keV [3]. However, in the case of non-metallic gratings the origin of the radiation must be understood in terms rather different to those used to describe the emission process from a 'perfect' metallic conductor. Instead of a 'sea' of free electrons on the surface of the grating that can respond instantaneously to the electron's field, one might now think in terms of radiating dipoles that are induced inside the material by that field. The brief discussion that follows is not very rigorous but it does provide the physical framework for this process. It is based closely on Jackson's treatment of transition radiation [4], where the reader can find a more detailed and rigorous analysis and further references. The mathematics have been kept to a minimal level but sufficient to allow the user to implement an analysis code.

Consider a slab of semiconductor material, say silicon, which we assume for convenience to be infinite in the y-direction, in analogy with the 'infinite width' metallic grating considered in Chapter 1, and an electron travelling in the z-direction with velocity v and at a height x_0 above the silicon surface (see the schematic of fig. 7.2). We want to calculate the spectral density of the radiation that is due to the varying polarization of the material, at an observation point P that is at a distance r from the origin O.

It is also assumed that at the rather high frequencies that are relevant to the present discussion (~3 × 10^{14} Hz), the index of refraction is of order unity and that the field of the electron inside the silicon will be about the same as its value in vacuum. Therefore, the field at a point $S(z,\psi)$ inside the semiconductor, which is at a distance s from the origin, will be:

$$\overline{E} = \overline{E}_0 e^{\frac{i\omega z}{v}}$$

where $\overline{E}_0$ is the field incident at the origin O; the angle ψ is used to define the azimuthal position of the element S inside the material. The field $\bar{E}$ will induce a time-dependant polarization $\bar{P}$ in the elementary volume d^3s around S:

$$\overline{P} = \frac{\varepsilon(\omega) - 1}{4\pi}\bar{E}$$

The induced dipole radiation field is given by:

$$d\overline{E}_{rad} = \frac{e^{ikR}}{R}k^2\left(\overline{n} \times \overline{P}\right) \times \overline{n} d^3 s$$

where $\overline{k}$ is the wave vector and $\bar{n}$ is the unit vector in the direction of observation defined by:

$$\overline{n} = (\sin\theta cos\phi,\ \ \sin\theta\sin\phi,\ \ \cos\theta)$$

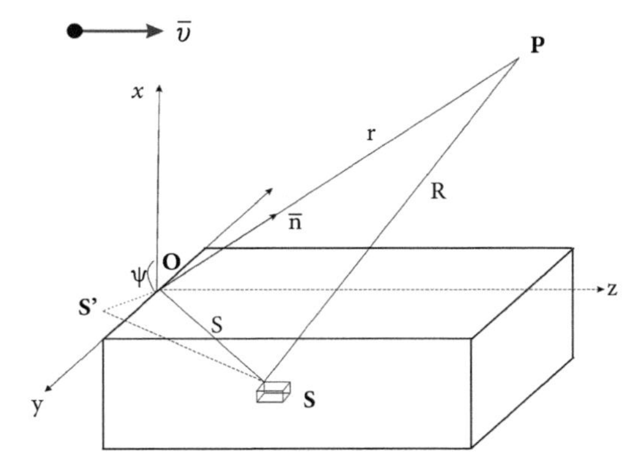

Fig. 7.2 Schematic used for the calculation of the spectral density of the radiation at point P.

The polarization is proportional to the field incident at S, and since $R \cong r - \bar{k}.\bar{s}$:

$$\bar{E}_{rad} \cong \frac{e^{ikr}}{r} k^2 \frac{\varepsilon(\omega) - 1}{4\pi} \int (\bar{n} \times \bar{E}) \times \bar{n} e^{-i\bar{k}.\bar{s}} d^3 s$$

Assuming that the dielectric constant at these high frequencies can be approximated by the expression:

$$\varepsilon(\omega) \simeq 1 - \frac{\omega_p^2}{\omega^2}$$

where ω_p is the plasma frequency of the material, then the previous expression for the radiated field becomes:

$$\bar{E}_{rad} \cong -\frac{e^{ikr}}{r} \frac{k^2}{4\pi} \frac{\omega_p^2}{\omega^2} \int (\bar{n} \times \bar{E}) \times \bar{n} e^{-i\bar{k}.\bar{s}} d^3 s$$

In terms of the field $\bar{E}_0(\omega)$ incident at the origin O, the spectral density of the emitted radiation is then given by:

$$\frac{d^2 I}{d\omega d\Omega} = \frac{\omega_p^4}{32\pi^3 c^3} \left| \int (\bar{n} \times \bar{E}_0) \times \bar{n} e^{-ik_x x} e^{-ik_y y} e^{i\left(\frac{\omega}{v} - k_z\right) z} dx dy dz \right|^2 \tag{7.1}$$

We use the symbol $\bar{G}$ to denote the integral inside the squared modulus of eq. 7.1. Integration with respect to z is straightforward and assuming that the length (l) of the semiconductor slab is not very short, we obtain:

$$\bar{G} \cong \frac{i}{\left(\frac{\omega}{v} - k\cos\theta\right)} \int (\bar{n} \times \bar{E}_0) \times \bar{n} e^{-ik_x x} e^{-ik_y y} dx dy \tag{7.2}$$

Working in a way similar to that described in Chapter 1, the vector triple product in the integrand can be resolved along two directions with unit vectors $\bar{\varepsilon}_1, \bar{\varepsilon}_2$ (see fig. 1.8), with $\bar{\varepsilon}_1 = (\cos\theta\cos\phi, \cos\theta\sin\phi, -\sin\theta)$ and $\bar{\varepsilon}_2 = (-\sin\phi, \cos\phi, 0)$. The projections of the vector product in $\bar{G}$ on these two directions are:

$$\bar{E}.\bar{\varepsilon}_1 = E_x \cos\theta\cos\phi + E_y \cos\theta\sin\phi - E_z \sin\theta = E_\rho \cos\psi\cos\theta\cos\phi + E_\rho \sin\psi\cos\theta\sin\phi - E_z \sin\theta$$

and

$$\bar{E}.\bar{\varepsilon}_2 = -E_x \sin\phi + E_y \cos\phi = -E_\rho \cos\psi\sin\phi + E_\rho \sin\psi\cos\phi$$

where E_ρ and E_z are the Fourier transforms of the radial and longitudinal components of the electron's field, respectively. The expressions for the field components are:

$$E_\rho = \sqrt{\frac{2}{\pi}\frac{e\omega}{\gamma v^2}} K_1\left(\frac{\omega\sqrt{x^2+y^2}}{v\gamma}\right)$$

$$E_z = -i\sqrt{\frac{2}{\pi}\frac{e\omega}{\gamma^2 v^2}} K_0\left(\frac{\omega\sqrt{x^2+y^2}}{v\gamma}\right)$$

where K_1 and K_0 are the modified Bessel functions and the other symbols have their usual meaning. Therefore, the integral (2) can be broken down into a number of integrals of the type:

$$G \sim \int_a^b e^{-ik_x x} dx \int_{-\infty}^{\infty} \frac{\cos k_y y}{\sqrt{x^2+y^2}} K_1\left(\frac{\omega\sqrt{x^2+y^2}}{v\gamma}\right) dy$$

or

$$G \sim \int_a^b e^{-ik_x x} dx \int_{-\infty}^{\infty} \frac{y \sin k_y y}{\sqrt{x^2+y^2}} K_1\left(\frac{\omega\sqrt{x^2+y^2}}{v\gamma}\right) dy$$

or

$$G \sim \int_a^b e^{-ik_x x} dx \int_{-\infty}^{\infty} \cos k_y y K_0\left(\frac{\omega\sqrt{x^2+y^2}}{v\gamma}\right) dy$$

These are all integrals that are familiar from the treatment of the metallic grating in Chapter 1. If the assumption of infinite width is valid, then there are analytical expressions for the y-integration. The limits of the x-integrals will depend on x_0 and on the thickness of the slab but the x-integration will have to be done numerically. It is, therefore, possible to construct a semi-analytical code in order to predict the SP radiation yield from a silicon micro-grating, operating in the micron or sub-micron region of the spectrum.

One might well ask to what an extent is this approach valid? The answer will depend on the validity of the initial assumption that the index of refraction, at the wavelengths of interest here, is of order unity. The subject of the frequency dependence of the complex dielectric constant $\tilde{\varepsilon}$ (or of the complex index of refraction $\tilde{n}$) of a material is dealt with in many textbooks [5, 6]. For silicon, which is the material assumed in the above sample calculation, the data reported in [7] indicate that at a wavelength of 0.82 μm ($E \cong 1.5eV$) the refractive index for crystalline silicon is about 3.7; thereafter, it decreases down to about unity but at wavelengths shorter than those under consideration here ($\lambda = 0.21\,\mu m$ or $E = 6\,eV$). Therefore, the above treatment of the emission process from a silicon grating might not be appropriate and the use of a numerical simulation package for the calculation of the emitted intensity would be preferable. In actual practice, however, a silicon nanograting is more likely to be fabricated from lightly doped silicon [3] in order to make the sample conductive and to avoid the inevitable charging of any insulating material which is brought close to the electron beam. In this case, the optical properties of the *doped* material are the ones that are really relevant and these could be significantly different from those of a pure, crystalline sample [8, 9]. Recent work suggests that the index of refraction for low-resistivity silicon wafers decreases to about 2 for frequencies in the 6 THz region [9], but one would need to know the optical properties of the material used in the manufacture of the silicon grating before deciding on the validity, or otherwise, of the above approximation.

References

1. A. Massuda et al, ACS Photonics **5** (2018) 3513
2. Y. Yang et al, Nature Physics Letters **14** (2018) 894
3. C. Roques-Carmes et al, Nature Communications (2019) https://doi.org/101038/s41467-019-11070-7
4. J.D. Jackson, 'Classical Electrodynamics', John Wiley & Sons, 2nd Edition, Ch. 14, Section 9 (1975)
5. M. Born and E. Wolf, Principles of Optics, Ch. XIII, Cambridge University Press (2019)
6. C. Kittel, 'Introduction to Solid State Physics', John Wiley & Sons, 3rd Edition, Ch. 8 (1967)
7. D.E. Aspnes and A.A. Studna, Phys. Rev. B **27** (1983) 985
8. S. Basu, B.J. Lee, and Z.M. Zhang, Journal of Heat Transfer **132** (2010) 023302
9. N. Chudpooti et al, Mater. Res. Express **8** (2021) 056201

Epilogue

Any application of the Smith–Purcell (SP) radiative process, whether as a diagnostic tool for the properties of the charged particle (usually electron) beam that gives rise to it or as a tuneable source of radiation, will require a method for calculating the expected yield. The emphasis in this text has been on providing the reader with a clear physical picture of the origins of the radiation, based on the acceleration of the surface charges on the grating, together with an outline of the mathematical methods required for a semi-analytical calculation of the radiated power, under different experimental circumstances. This should allow the experimenter to adjust the design of the grating in order to achieve maximum yield at the wavelengths of interest. Calculations and experimental measurements of the emitted radiation are in good agreement, over a very wide range of beam energies, from a few MeV to 28 GeV, but there is scope for improvement both on the theoretical and on the experimental side. Since background radiation, i.e. non-SP radiation, is likely to be an important issue in any accelerator-based application, an effort has been made to discuss ways of suppressing the background or distinguishing it from the desired SP signal. Accurate assessment of the true SP radiation signal arriving at a given detector position, together with an accurate calibration of the detector responsivity at that wavelength, are essential requirements for the correct reconstruction of the time profile of the charged particle bunch. As far as the use of SP radiation as a tuneable source of radiation is concerned, the very brief overview of the existing sources (Chapter 5), together with the more extensive discussion of wavelength coverage and methods for selecting a wavelength from an SP radiation source (Chapters 6 and 7), will allow the experimenter to assess the relative advantages of SP-based sources.

Finally, it is worth pointing out the educational aspect of research in this field. As has been stated right at the beginning, SP radiation is a phenomenon of classical electromagnetism but its study involves not only electromagnetism but most of classical physics, such as optics, solid state physics, etc., plus some special relativity. At a time when accelerator-based experiments are inevitably becoming more and more complicated, research on SP radiation is one of the few areas where it is still possible for the student or the researcher to be intimately involved with all aspects of the work, 'from the beginning to the end': the creation of the electron beam, the evaluation of its properties, the beam transport to the target, the design of the optical system for collecting the emitted radiation, filters, detectors, etc., and finally the analysis of the results, all these could be within the grasp of one individual. Purcell's two statements, one in his original paper ('this effect may have interesting applications') [1] and the other in a private communication ('good for teaching physics') [2], are both correct.

References

1. S.J. Smith and E.M. Purcell, Phys.Rev. **92** (1953) 1069
2. E.M. Purcell, private communication, 1992

Appendix

Instrumentation

It is useful at this point to include a few general observations and comments about the instrumentation that would be required in any application of the Smith–Purcell (SP) radiation, especially in the THz region. Since the discussion on the sources of THz radiation has been presented in a previous chapter, we restrict the discussion here on the topics of detectors and filters. This is, again, a mature field of research and development that predates any work on SP radiation. The purpose of the discussion is to alert the reader as to what instruments are available and the basic physical principles of their operation. Further details can be found in the cited references and in particular in references [1,2]. The first is rather old now but still very readable, whereas the second is much more recent and comprehensive.

A.1 Detectors

The detectors used in this part of the spectrum are based on one of the following physical processes: (a) the THz radiation heats up the body of the detector and this causes some physical change which can be measured and converted into a signal; these are the thermal detectors. In the Golay cell, for example, the heating causes the deformation of a small volume of gas, while in another group of thermal detectors called 'bolometers' the heat causes a change in the electrical resistance, which again can be measured. In the same category belong also the pyroelectric detectors, where the heat of the radiation causes a change in the polarization of the material. (b) The incoming radiation causes some electronic transition in the material, say from the valence to the conduction band or from a donor level to the conduction band and this is again converted into an output signal. These are known as electronic or photoconductive detectors. There are other types of electronic transition that can be induced and hence various types of detector, one particular example being the Schottky barrier detector where the photon energy is sufficient to promote electrons over the potential barrier of the metal–semiconductor interface. (c) The radiation signal is mixed with a local oscillator signal of approximately the same frequency and another detector measures the strength of the signal corresponding to the difference of the two frequencies. It should be emphasized that the above classification is not very precise and that there are detectors that may not fit neatly into one of the above categories, one notable case being that of the hot electron bolometer where the absorbed radiation heats the conduction band electrons, rather than the lattice itself.

All detectors are characterized by parameters that essentially describe their quality and their performance. The most important of these is the ability of the detector to detect signals over a wide range of THz frequencies; this is often referred to as the spectral range or spectral or frequency bandwidth. A wide spectral range is obviously desirable. Another important parameter is the noise equivalent power (NEP) which is a measure of the ability of the detector to discriminate against noise, which in this context means the unavoidable fluctuations in the output of any detector, even in the absence of an input signal. The sources of noise are varied, e.g. background radiation, noise from the electronics, random fluctuations in the signal itself, etc., but usually it is the background contribution that is the dominant one. The NEP is defined as the signal power required to give a signal-to-noise ratio of 1. The generation of noise is a random process and normally the longer a measurement lasts, the greater the accuracy in its determination. Therefore, in order to make valid comparisons between different detectors, it is important to specify the duration of the measurement or, equivalently, the frequency bandwidth of the measurement. By convention, the values of NEP are given in $W/\sqrt{Hz}$ but if an NEP value is given in W, the implication is that it has been taken with a noise bandwidth of 1 Hz. It is worth noting that according to the Nyquist theorem, a bandwidth of 1 Hz means a measurement duration of

0.5 s. The existence of the square root can be understood on the basis of the following argument based on probability and statistics: for a random process, the measured standard deviation of the mean, or standard error, (σ_m) is given by $\sigma_m = \frac{\sigma}{\sqrt{n}}$, where σ is the standard deviation of the parent distribution and n is the number of independent measurements of this quantity, which in our case is directly related to the duration of the measurement. In other words, the longer the measurement (or the narrower the bandwidth) the better the estimate of the noise. Sometimes the noise performance of a detector is given in terms of its detectivity, which is just the inverse of NEP.

The responsivity of the detector, which is defined as the change in output voltage (or current) per unit of signal power change and is expressed in dV/dW or dA/dW, is also important. If the responsivity is constant, i.e. it does not vary with the input power, the detector is 'linear'. This is a desirable characteristic, but not always achievable. It is also important to know the response time or modulation bandwidth of a detector since this is a measure of the ability of the detector to follow changes to the input signal. Apart from the above physical properties of far-infrared (FIR) detectors, one has to consider some practical issues, such as cost, size, robustness, operating temperature (does it need cooling?), and lifetime. In the context of SP work and in particular in the application of SP radiation to beam diagnostics, size and robustness are particularly important because the detectors are likely to be deployed in an accelerator beam line, where space is at a premium and where the environment can be hostile because of the high background radiation (X-rays and gamma rays). An extensive review article on THz detectors can be found in reference [3].

Some typical parameters for FIR detectors are listed in Table A.1, which has been extracted from a number of references. It should be emphasized that the list is far from comprehensive and the reader should consult reference [4] and the manufacturers' catalogues for a more thorough discussion of the subject.

The main conclusions from Table A.1 are: (a) thermal detectors offer the widest wavelength coverage while photoconductors are by their nature narrow band devices; (b) low temperature devices have lower NEP; and (c) Schottky barrier detectors are very fast.

Table A.1 Some FIR detectors and their properties

Detector type	Wavelength Range (mm)	NEP $W.Hz^{-0.5}$	Responsivity	Response time or modulation frequency	Operating temperature (K)	Ref.
Golay cell	4×10^{-4}-7.5	1.4×10^{-4}	1×10^5 V/W	30 ms	278–313	3
Pyroelectric	0.01–3	1×10^{-9}	7×10^4 V/W	5–30 Hz	267–393	3, 7
Hot electron bolometer (In Sb)	0.6–5	7.5×10^{-13}	3.5×10^3 V/W	1 MHz	4.2	6
Semiconductor bolometer (Si)	0.015–2	1.2×10^{-13}	2.5×10^5 V/W	200–400 Hz	4.2	3
Photoconductor (Ge:Ga)	0.06–0.2	8×10^{-13}	0.3 A/W	> 50 kHz	4.2	3, 6
Schottky barrier	0.17–0.27*	~1.1×10^{-10}	100 V/W	~40 GHz	~293	3, 5

*Other wavelength bands are available

A.2 Filters

It is difficult to overemphasize the importance of filtering in SP experiments, especially those that are accelerator based. The reason is that apart from the removal or suppression of wavelengths that lie outside the SP wavelength band, it is also important to account for wavelengths that lie within the SP band but whose origin is not the grating itself. This background radiation can be due to transition, diffraction, or synchrotron sources that may be quite some distance upstream from the location of the SP apparatus. As high vacuum apparatus tends to be formed from metal pipes, these can act as oversized waveguides, transferring unwanted radiation to the SPr detector.

Wavelengths that are outside the SP band can be suppressed by a combination of the available FIR filters. These fall into three categories: (a) long (wavelength) pass filters that cut off all the wavelengths below a certain value and allow through the long ones; (b) short pass filters that do exactly the opposite, i.e. have high transmission for wavelengths shorter than a certain cut-off value; and (c) band pass filters which have high transmission for a certain wavelength band. For the purposes of SP experiments, categories (a) and (b) are probably the more important. Particularly important and useful are the short-pass filters known as 'waveguide array plates' (WAP) filters [5]. These are formed by drilling a (usually) hexagonal pattern of holes on a metal plate. Each hole can be considered as a cylindrical waveguide, with a cut-off wavelength (λ_c) given by [6]:

$$\lambda_c = \frac{\pi d}{1.841}$$

where d is the hole diameter. This applies to an infinitely long waveguide but curtailment of the length by the finite thickness of the metal plate only affects the sharpness of the cut-off. There is a fair amount of literature, extending back over a number of years, about the design and properties of this type of filter. An example of such a filter is shown in the photograph of fig. A.1a. The filter was designed to be placed at an observation angle $\theta = 70^0$ relative to a grating having a period of 0.5 mm where the expected SP wavelength (λ_{SP}) is 329 μm. The cut-off wavelength was set at 1.2 times λ_{SP}, i.e. at 395 μm; hence, the required hole diameter, according to the above expression, is 230 μm. There are two other mechanical dimensions that are important in determining the transmission properties of the filter: (a) the thickness of the brass plate and (b) the spacing of the holes. These were set at 2.5 the hole diameter for the former and 1.345 the diameter for the latter, i.e. 575 μm and 310 μm, respectively. The filter consisted of 4,167 holes drilled in a hexagonal pattern on a 21 mm diameter brass disc. The measured transmission characteristics of the filter, which are independent of the polarization of the radiation, are shown in fig. A.1b. The peak transmission of the filter, about 50%, occurs beyond the target value, at just below 0.4 mm but

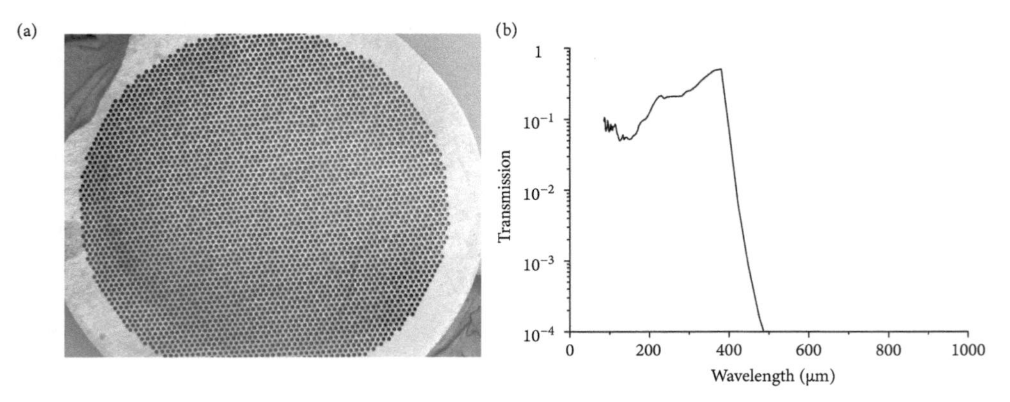

Fig. A.1 (a) A 'waveguide array plate' type of filter, designed for peak transmission at 329 μm; there are 4,167 holes with a diameter of 230 μm, drilled in a 575 μm thick, 21 mm diameter brass plate. (b) The measured transmission curve of this filter.

there is the expected steep fall in transmission beyond this peak. The drop in transmission at shorter wavelengths is attributed to diffraction effects. The performance of this filter is not perfect, probably due to manufacturing imperfections, but was deemed acceptable. A number of such filters covering the range 234 μm to 2650 μm were manufactured in the author's laboratory, all with roughly similar performance.

For even shorter cut-off wavelengths the manufacture of WAP filters becomes problematic. It is possible, however, to use electroformed metal meshes, as described in [7]. The transmission curve for such a filter is shown in fig. A.2. This particular mesh consisted of 77 μm diameter holes, on a 118 μm pitch and had a peak transmission of about 90% at about 122 μm, again independent of the polarization of the incident radiation.

For situations where a long pass filter is required, one can use polyethylene transmission gratings which can be made very simply by stamping crossed rectangular grooves on the two sides of a polyethylene disc. The transmission characteristics of these filters depend on the ratio of wavelength over grating pitch, as shown in fig. A.3. Further details about the manufacture and properties of these filters can be found in [1].

Before concluding this Appendix, it is useful to remind the reader that SP radiation is a rather weak signal and therefore particular attention should be paid to the light collection system so that the intensity that reaches the detector(s) can be maximized. Non-imaging light concentrators [8–10] (Winston cones) can be very useful in this respect. These are devices that can accept light that impinges on an entrance aperture, over a given angular range, and transmit it by multiple reflections on a suitably curved surface to an exit aperture, which is much smaller than the entrance one. The detector is placed behind the exit aperture. One consequence of this is that because of phase space conservation, the rays emerge from the exit aperture with a significant angular divergence and this must be considered when deciding the exact position of the FIR detector behind the exit aperture. They have been used extensively in solar energy applications and in particle physics [11]. A detailed description of Winston cones applied to SP radiation detection can be found in [12].

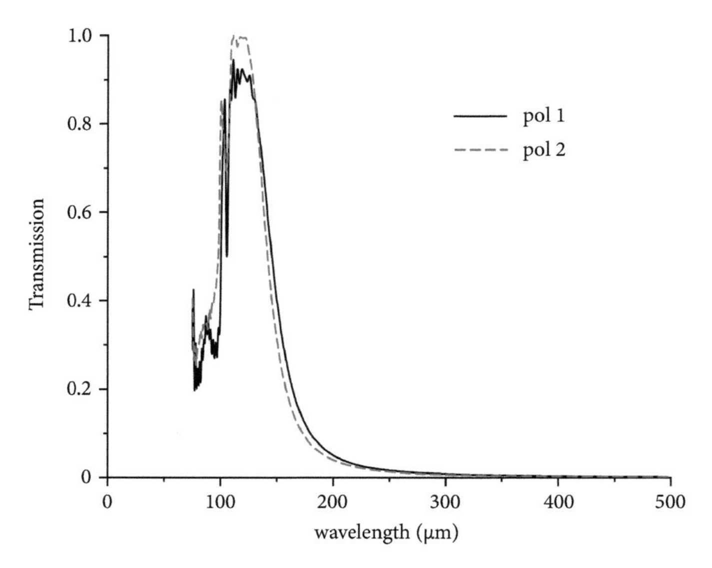

Fig. A.2 The transmission curves of an electroformed nickel mesh filter, for two polarization directions.

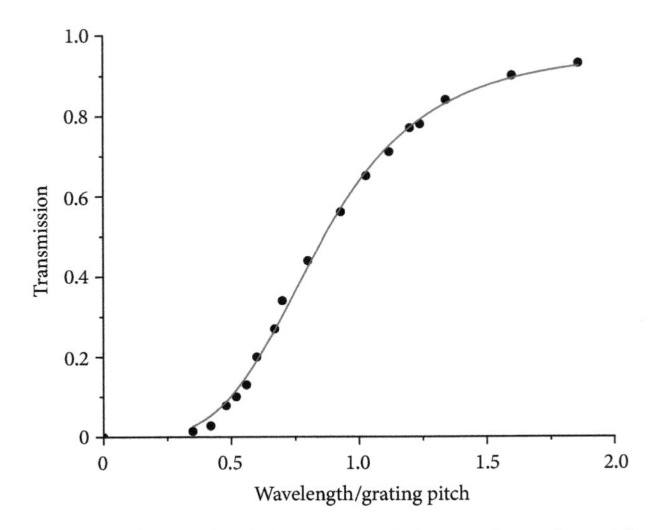

Fig. A.3 Transmission curve for a polyethylene transmission grating, adapted from ref. [1]. The curve is to guide the eye.

References

1. M.F. Kimmitt, 'Far-Infrared Techniques', Pion (1970)
2. E. Bründermann, H.W. Hubers, and M.F. Kimmitt, 'Terahertz Techniques', Springer (2012)
3. F. Sizov and A. Rogalski, Progress in Quantum Electronics **34** (2010) 278
4. R.A. Lewis, J. Phys. D: Appl. Phys. **52** (2019) 433001
5. C. Winnewisser, F. Lewen, and H. Helm, Appl. Phys. A **66** (1998) 593
6. C. Winnewisser et al, IEEE Trans. Microwave Theory & Techniques **48** (2000) 744
7. P.G. Huggard et al, Applied Optics **39** (1994) 39
8. W.T. Welford and R. Winston, 'Optics of non-imaging concentrators: Light and Solar energy', Academic Press (1978)
9. W.T. Welford and R. Winston, 'High collection non-imaging optics', Academic Press (1990)
10. R. Winston, 'Non-imaging Optics' Sci. Amer. **264** (1991) 76
11. G. Doucas et al, Nucl. Instr. and Methods in Phys. Res. A **370** (1996) 579
12. V.J. Blackmore, D. Phil. Thesis, University of Oxford (2008) unpublished

Index